T0214383

# SpringerBriefs in Molecular Science

## Biobased Polymers

**Series editor**

Patrick Navard, CNRS/Mines ParisTech, Sophia Antipolis, France

Published under the auspices of EPNOE*Springerbriefs in Biobased polymers covers all aspects of biobased polymer science, from the basis of this field starting from the living species in which they are synthetized (such as genetics, agronomy, plant biology) to the many applications they are used in (such as food, feed, engineering, construction, health, …) through to isolation and characterization, biosynthesis, biodegradation, chemical modifications, physical, chemical, mechanical and structural characterizations or biomimetic applications. All biobased polymers in all application sectors are welcome, either those produced in living species (like polysaccharides, proteins, lignin, …) or those that are rebuilt by chemists as in the case of many bioplastics.

Under the editorship of Patrick Navard and a panel of experts, the series will include contributions from many of the world's most authoritative biobased polymer scientists and professionals. Readers will gain an understanding of how given biobased polymers are made and what they can be used for. They will also be able to widen their knowledge and find new opportunities due to the multidisciplinary contributions.

This series is aimed at advanced undergraduates, academic and industrial researchers and professionals studying or using biobased polymers. Each brief will bear a general introduction enabling any reader to understand its topic.

*EPNOE The European Polysaccharide Network of Excellence (www.epnoe.eu) is a research and education network connecting academic, research institutions and companies focusing on polysaccharides and polysaccharide-related research and business.

More information about this series at http://www.springer.com/series/15056

Merin Sara Thomas · Rekha Rose Koshy
Siji K. Mary · Sabu Thomas
Laly A. Pothan

# Starch, Chitin and Chitosan Based Composites and Nanocomposites

Springer

Merin Sara Thomas
Department of Chemistry
C.M.S. College
Kottayam, Kerala, India

and

Department of Chemistry
Mar Thoma College
Thiruvalla, Kerala, India

Rekha Rose Koshy
Department of Chemistry
C.M.S. College
Kottayam, Kerala, India

and

Department of Chemistry
Bishop Moore College
Mavelikara, Kerala, India

Siji K. Mary
Department of Chemistry
C.M.S. College
Kottayam, Kerala, India

and

Department of Chemistry
Bishop Moore College
Mavelikara, Kerala, India

Sabu Thomas
International and Interuniversity Centre for
Nanoscience and Nanotechnology
Mahatma Gandhi University
Kottayam, Kerala, India

Laly A. Pothan
Department of Chemistry
C.M.S. College
Kottayam, Kerala, India

and

Department of Chemistry
Bishop Moore College
Mavelikara, Kerala, India

ISSN 2191-5407    ISSN 2191-5415 (electronic)
SpringerBriefs in Molecular Science
ISSN 2510-3407    ISSN 2510-3415 (electronic)
Biobased Polymers
ISBN 978-3-030-03157-2  ISBN 978-3-030-03158-9 (eBook)
https://doi.org/10.1007/978-3-030-03158-9

Library of Congress Control Number: 2018960176

This Springer imprint is published by the registered company Springer Nature Switzerland AG
The registered company address is: Gewerbestrasse 11, 6330 Cham, Switzerland

# Contents

# Abbreviations

| | |
|---|---|
| AFM | Atomic force microscopy |
| AuNPs | Gold nanoparticles |
| Au@CDs-CS/GCE | Gold@carbon dots-chitosan-modified glassy carbon electrode |
| B.mori | Bombyx mori |
| CA | Citric acid |
| CCNGs | Curcumin-loaded chitin nanogels |
| CF | Coir fiber |
| CHNC | Chitin nanocrystals |
| CHNF | Chitin nanofibers |
| ChOx | Cholesterol oxidase |
| CMC | Carboxymethyl cellulose |
| CNT | Carbon nanotube |
| CNF | Cellulose nanofiber |
| CNP | Chitin nanoparticle |
| CNW | Chitin nanowhiskers |
| Cr(VI) | Chromium |
| CS | Chitosan |
| CZB | Chitosan hydrogel/nano zinc oxide composite bandages |
| α-chitin | Alpha chitin |
| β-chitin | Beta chitin |
| °C | Degree Celsius |
| DMA | Dynamic mechanical analysis |
| DSC | Differential scanning calorimetry |
| dTG | Derivative thermogram |
| ECM | Extracellular matrix |
| E.coli | Escherichia coli |
| EM | Elastic modulus |
| ENR | Epoxidized natural rubber |
| $Fe_3O_4$ | Magnetite |

| | |
|---|---|
| $\gamma$-Fe$_2$O$_3$ | Maghemite |
| GeO$_2$ | Germanium dioxide |
| GPa | Giga pascal |
| GPS | Glycerol-plasticized potato starch |
| H | Gelatinization enthalpy |
| HA | Hydroxyapatite |
| ITO | Indium tin oxide |
| L. monocytogenes | Listeria monocytogenes |
| MC3T3-E1 | Mouse osteoblastic cells |
| MgG | Magnesium gluconate |
| MgO | Magnesium oxide |
| MMT | Montmorillonite |
| MPa | Mega pascal |
| MTT | 3-(4,5-dimethylthiazol-2-yl)-2,5-diphenyltetrazolium bromide |
| MWCNT | Multi-walled carbon nanotubes |
| $\mu$m | Micrometer |
| N-chitin | Nano chitin gel |
| NFC | Nanofibrillar cellulose |
| NIH-3T3 | Mouse embryo fibroblast cell line |
| nm | Nanometer |
| NP | Nanoparticles |
| OPEFB | Oil palm empty fruit bunch |
| PCC | Chitin-coated with polyaniline |
| PCL | Polycaprolactone |
| PEGDA | Polyethylene glycol diacrylate |
| PEO | Poly(ethylene oxide) |
| PHBV | Poly(3-hydroxybutyrate-co-3-hydroxyvalerate |
| PHF | Pea hull fiber |
| PHFNW | Cellulose nanowhiskers from pea hull fiber |
| PLA | Polylactic acid |
| PLGA | Poly(lactic-co-glycolic acid) |
| PMMA | Polymethylmethacrylte |
| PS/GO-n | Glycerol-plasticized pea starch/graphene oxide |
| PVA | Polyvinylalcohol |
| PVP | Poly(N-vinylpyrrolidone) |
| RCB | Bleached rice hull |
| RH | Rice hull |
| RHNF | Cellulose nanofibrils extracted from rice hull |
| S/CHNC | Starch/Chitin nanocrystal |
| S/CHNF | Chitin nanofiber-reinforced starch matrix |
| SCB | Sugarcane baggase |
| SEM | Scanning electron microscope |
| ST | Starch |
| SiO$_2$ | Silicon dioxide |

| | |
|---|---|
| SPI | Soy protein isolate |
| $T_c$ | Crystallization temperature |
| TEC | Tri Ethyl Citrate |
| TEOG | Germanium tetraethoxide |
| TEM | Transmission electron microscope |
| TEMPO | 2, 2, 6, 6-tetramethyl-piperidinyl-1-oxyl |
| TGA | Thermogravimetric analysis |
| $TiO_2$ | Titanium dioxide |
| $T_o$ | Onset temperature |
| $T_p$ | Peak temperature |
| TPS | Thermoplastic starch |
| WPI | Whey protein isolate |
| wt% | Weight percentage |
| WVP | Water vapour permeability |
| WVTR | Water vapour transmission rate |
| w/w% | Weight/weight percent |
| ZnO | Zinc oxide |
| $ZrO_2$ | Zirconium dioxide |

# Chapter 1
# Introduction

Concern about the depletion of natural resources and environment pollution, caused by petroleum-based plastics, has drawn attention to the development of environment friendly polymer composites and nanocomposites, for applications in food, cosmetics, and pharmaceutical industries. Natural polymers are much more attractive than artificial polymers for the preparation of composites, due to their "green" characteristics, such as biodegradability, biocompatibility, renewability, and sustainability (Lu et al. 2013). Biopolymers, like polysaccharides (starch, cellulose, chitin and chitosan) and proteins (soy protein, wheat protein, casein, and gelatin) obtained from the nature, are the most viable alternative for producing green materials in the near future. Polysaccharides, a class of natural macromolecules, have the tendency to be extremely bioactive, and are generally derived using different biotechnological approaches from agricultural feedstock or crustacean shell wastes (Anitha et al. 2014).

Naturally occurring polymers perform a diverse set of functions in their native setting. Polysaccharides function in membranes and intracellular communication, in recognition events at the cell surface, as cell wall structures, as capsular layers or protective barriers around cells, as emulsifiers, as adhesives, and as sequestering agents for water, nutrients and metals for cells. The benefits of using naturally occurring polymers for material applications are many. For example, environmental compatibility would be enhanced since no environmental burdens would be introduced due to their use. In addition, the utilization of renewable resources provides an incentive to extend nonrenewable petrochemical supplies (Kaplan et al. 1998). Here we are discussing the use of starch, chitin and chitosan as fillers in polymer composites and nanocomposites.

## 1.1   Starch

Polysaccharides are good candidates for biodegradable films. Starch is considered as the most promising among them due to its abundance, low cost, biodegradability and renewability. Starch based composites and nanocomposites are now considered to be next generation "green" materials, because of its capability to be converted into thermoplastic materials (Tang and Alavi 2011; Gironès et al. 2012; Lopez et al. 2014). In biocomposites, starch can be used either as the continuous polymeric phase (matrix) or the dispersed phase (filler) or both. Nano reinforced starch based composites generally exhibit enhanced mechanical and thermal properties when nanofillers are well dispersed (Yu et al. 2009; Kaushik et al. 2010; da Silva et al. 2015; Li et al. 2015; González et al. 2015; Cao et al. 2008; Salehudin et al. 2014; Tang and Alavi 2011).

Starch consists of linear polysaccharide amylose and highly branched polysaccharide (Fig. 1.1). Depending on the source, starch generally contains 20–25% amylose and 75–80% amylopectin. Amylose is a semicrystalline biopolymer and is soluble in hot water, while amylopectin is highly crystalline and is insoluble in hot water.

Starch based biodegradable composites have drawn considerable attention of the scientific community over the last two decades due to their potential applications in many industries including food and medicine. However, in comparison with conventional synthetic materials, starch based biodegradable products exhibit many disadvantages such as poor resistance to water, brittleness and poor elasticity. All of these disadvantages can be attributed to the highly hydrophilic characteristics of starch. One of the effective methods to improve the above mentioned detriments is to incorporate various fillers. The use of natural fibers and nano reinforcements for the elaboration of composites and nanocomposites is an effective way to improve properties of thermoplastic starch (Babu et al. 2013; Orue et al. 2014).

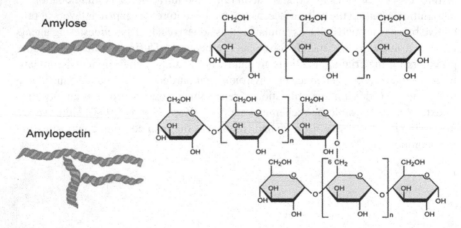

**Fig. 1.1** Structure of amylose and amylopectin in starch. Reprinted with permission from Zia et al. (2016). Copyright Elsevier

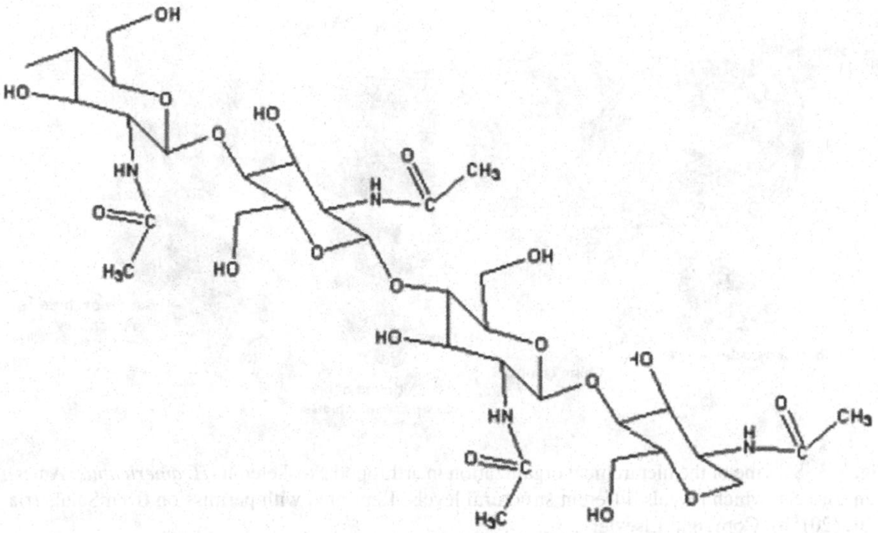

**Fig. 1.2** Molecular structure of chitin. Reprinted with permission from Visakh and Thomas (2010). Copyright Springer

Polysaccharides based on chitin and chitosan have received much attention in recent years as new functional biomaterials with potential applications in various fields. Chitin is the second most ubiquitous natural polysaccharide after cellulose on earth and is composed of $\beta(1 \rightarrow 4)$-linked 2-acetamido-2-deoxy-$\beta$- D-glucose (*N*-acetylglucosamine). It has acetamide groups ($-NHCOCH_3$) at the C2 positions (Fig. 1.2).

Chitins are present as the main components in crab and shrimp shells, in the outer skins or cuticles of other arthropods, and in the molluscan shell of squid (so-called squid pen), coexisting with proteins and certain minerals (Boonurapeepinyo et al. 2011; Tzoumaki et al. 2011; Ifuku and Saimoto 2012). Figure 1.3 shows the scheme of the hierarchical organization in arthropod exoskeleton (*H. americanus*, American lobster), which reveals different structural levels.

Two types of chitin crystal are known, $\alpha$- and $\beta$-chitins. Most natural chitins have the $\alpha$-type crystal structure, while the $\beta$-type chitin is present in squid pens and tube-worms. In $\alpha$-chitins, all molecular chains are arranged in an antiparallel mode with strong intermolecular hydrogen bonding. On the other hand, $\beta$-chitins have a parallel chain packing mode. Especially, the intermolecular forces in squid pen $\beta$-chitins are weaker than those in $\alpha$-chitins, and this makes squid pen $\beta$-chitins more susceptible to enzymatic degradations or chemical reactions. The lateral dimensions of the crystalline fibrils of chitins range from 2.5 to 25 nm, depending on their biological origins (Fan et al. 2008b). Most of the naturally occurring polysaccharides e.g. cellulose, dextrin, pectin, alginic acid, agar, agarose and carrageenans are neutral or acidic in nature, whereas, chitin is an example of highly basic polysaccharide. Their unique properties include optical structural characteristics, ability to form films and

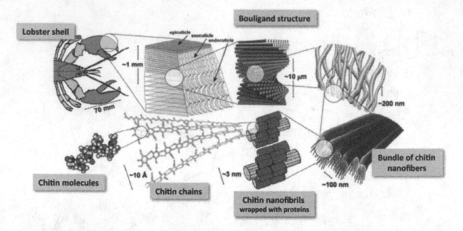

**Fig. 1.3** Scheme of the hierarchical organization in arthropod exoskeleton (*H. americanus*, American lobster), which reveals different structural levels. Reprinted with permission from Salaberria et al. (2015b). Copyright Elsevier

chelate metal ions (Visakh and Thomas 2010). Chitin has great potential to be used in the fabrication of implant devices, wound dressing materials, drug delivery systems, and regenerative medical components for bones and other medical materials, due to its high crystallinity, high strength, and biocompatibility (Fan et al. 2008b). Commercially chitin and chitosan are of great importance owing to their relatively high percentage of nitrogen (6.89%) compared to synthetically substituted cellulose. The following three steps in chronological order of the process are needed to produce chitin from crustacean shells: deproteinization (i) removal of residual proteins by chemical (NaOH) or enzymatic hydrolysis; (ii) demineralization—removal of mineral salts by acid treatment; and finally (iii) removal of lipids and pigments by typical bleaching treatments. In some cases, when the raw material is rich in minerals, it is preferable that the demineralization operation precedes the deproteinization process. After this, purified chitin could be: (i) dried and cracked into powders or small flakes; or (ii) kept wet in suspension. Figure 1.4 shows a schematic illustration of the conventional process of the chitin isolation from shell wastes.

Chitin nano-objects can be obtained by two approaches: top-down and bottom-up. As chitin fibrils are composed of two regions, i.e. crystalline and amorphous, chitin can be turned in nanocrystals, nanofibers and nanowhiskers via top-down method. This approach breaks down the chitin fibrils from native chitin into nanofibrils. Acid hydrolysis (Fan et al. 2008a), 2,2,6,6-tetramethyl-piperidinyl-1-oxyl (TEMPO)-mediated oxidation (Fan et al. 2008b), grinding and high-pressure homogenizing (Lu et al. 2013) are some representative techniques of this approach. The isolation of these nano-size chitin using: (i) acid conditions results in chitin nanocrystals (CHNC), which are rod like in appearance with low aspect ratio, (ii) mechanical treatments/disintegration results in chitin nanofibers (CHNF), which are fibrillar in appearance with high aspect ratio, and have lower crystallinity than CHNC. Chitin

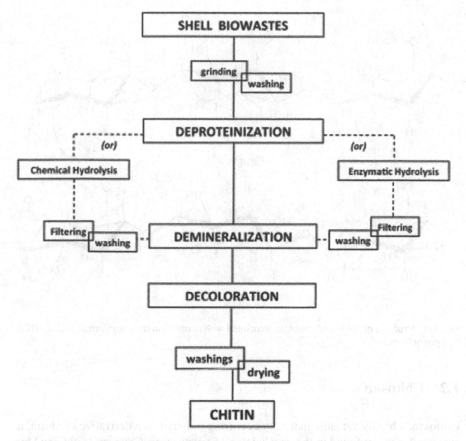

**Fig. 1.4** Illustrative scheme of the extraction process of chitin from shell wastes. Reprinted with permission from Salaberria et al. (2015a). Copyright Elsevier

nano-whiskers (CNW) of slender parallelepiped rods have been successfully prepared from chitin, using dilute acid hydrolysis at high temperature, followed by repeated mechanical treatment. This has been recently explored in nanotechnology applications (Mincea et al. 2012; Muzzarelli 2011). CNWs have many excellent properties, including biodegradability, biocompatibility, renewability, and antibacterial properties. On the other hand, self-assembled chitin nano-objects have been produced by regeneration from chitin solutions or gels via the bottom-up approach.

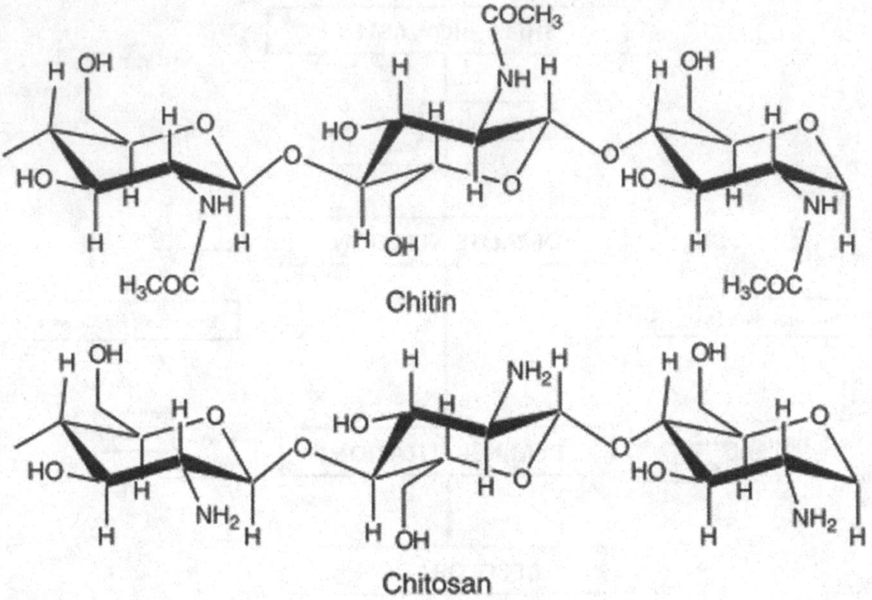

Chitin

Chitosan

**Fig. 1.5** Structure of chitin and chitosan. Reprinted with permission from Jayakumar et al. (2010). Copyright Elsevier

## 1.2   Chitosan

Chitosan, a highly versatile naturally occurring polymer, is a derivative of chitin, a structural element found in the exoskeleton of crustaceans. Chitosan is obtained by the deacetylation of chitin. Figure 1.5 shows the structure of chitin and chitosan. A higher degree of deacetylation corresponds to a higher percentage of positively charged primary amines and an overall higher charge density. Commercially available chitosan results from alkaline deacetylation of chitin. It is an attractive biocompatible, biodegradable and nontoxic natural biopolymer that exhibits excellent film-forming ability. Chemically, chitosan is a linear polysaccharide composed of glucosamine and N-acetyl glucosamine units linked by β(1-4) glycosidic bonds (Levengood and Zhang 2014). There are many forms of pure chitosan, which differ by their degrees of deacetylation (DD) and molecular weights. The degree of deacetylation represents the glucosamine to N-acetyl-glucosamine ratio. Both the degree of deacetylation and the molecular weight have a strong influence on other physicochemical properties of chitosan including crystallinity, solubility, and degradation (Levengood and Zhang 2014). An important property of chitosan is the ease with which it can be functionalized due to the presence of reactive primary amines and primary and secondary hydroxyl groups.

# References

Anitha A, Sowmya S, Kumar PTS, Deepthi S, Chennazhi KP, Ehrlich H, Tsurkan M, Jayakumar R (2014) Chitin and chitosan in selected biomedical applications. Prog Polym Sci 39:1644–1667. https://doi.org/10.1016/j.progpolymsci.2014.02.008

Babu RP, O'Connor K, Seeram R (2013) Current progress on bio-based polymers and their future trends. Prog Biomater 2:8. https://doi.org/10.1186/2194-0517-2-8

Boonurapeepinyo S, Jearanaikoon N, Sakkayawong N (2011) Reactive Red (RR141) solution adsorption by nanochitin particle via XAS and ATR-FTIR techniques. Int Trans J Eng Manag Appl Sci Technol 2:461–470

Cao X, Chen Y, Chang PR, Muir AD, Falk G (2008) Starch-based nanocomposites reinforced with flax cellulose nanocrystals. Express Polym Lett 2:502–510. https://doi.org/10.3144/expresspolymlett.2008.60

da Silva JBA, Nascimento T, Costa LAS, Pereira FV, Machado BA, Gomes GV, Druzian JI (2015) Effect of Source and Interaction with Nanocellulose Cassava Starch, Glycerol and the properties of films bionanocomposites. Mater Today Proc 2:200–207. https://doi.org/10.1016/j.matpr.2015.04.022

Fan Y, Saito T, Isogai A (2008a) Chitin nanocrystals prepared by TEMPO-mediated oxidation of α-chitin. Biomacromol 9:192–198. https://doi.org/10.1021/bm700966g

Fan Y, Saito T, Isogai A (2008b) Preparation of chitin nanofibers from squid Pen β-chitin by simple mechanical treatment under acid conditions. Biomacromol 9:1919–1923. https://doi.org/10.1021/bm800178b

Gironès J, López JP, Mutjé P, Carvalho AJFD, Curvelo AADS, Vilaseca F (2012) Natural fiber-reinforced thermoplastic starch composites obtained by melt processing. Compos Sci Technol 72:858–863. https://doi.org/10.1016/j.compscitech.2012.02.019

González K, Retegi A, González A, Eceiza A, Gabilondo N (2015) Starch and cellulose nanocrystals together into thermoplastic starch bionanocomposites. Carbohydr Polym 117:83–90. https://doi.org/10.1016/j.carbpol.2014.09.055

Ifuku S, Saimoto H (2012) Chitin nanofibers: preparations, modifications, and applications. Nanoscale 4:3308. https://doi.org/10.1039/c2nr30383c

Jayakumar R, Prabaharan M, Nair SV, Tamura H (2010) Novel chitin and chitosan nanofibers in biomedical applications. Biotechnol Adv 28(1):142–150. https://doi.org/10.1016/j.biotechadv.2009.11.001

Kaplan DL (1998) Introduction to biopolymers from renewable resources. Biopolymers from renewable resources. Springer, Berlin Heidelberg, pp 1–29

Kaushik A, Singh M, Verma G (2010) Green nanocomposites based on thermoplastic starch and steam exploded cellulose nanofibrils from wheat straw. Carbohydr Polym 82:337–345. https://doi.org/10.1016/j.carbpol.2010.04.063

Levengood SL, Zhang M (2014) Chitosan-based scaffolds for bone tissue engineering. J Mater Chem B 2(21):3161. https://doi.org/10.1039/C4TB00027G

Li M, Li D, Wang L, Adhikari B (2015) Creep behavior of starch-based nanocomposite films with cellulose nanofibrils. Carbohydr Polym 117:957–963. https://doi.org/10.1016/j.carbpol.2014.10.023

Lopez O, Garcia M, Villar M, Gentili A, Rodriguez MS, Albertengo L (2014) Thermo-compression of biodegradable thermoplastic corn starch films containing chitin and chitosan. LWT Food Sci Technol 57:106–115. https://doi.org/10.1016/j.lwt.2014.01.024

Lu Y, Sun Q, She X, Xia Y, Liu Y, Li J, Yang D (2013) Fabrication and characterisation of α-chitin nanofibers and highly transparent chitin films by pulsed ultrasonication. Carbohydr Polym 98:1497–1504. https://doi.org/10.1016/j.carbpol.2013.07.038

Mincea M, Negrulescu A, Ostafe V (2012) Preparation, modification, and applications of chitin nanowhiskers: a review. Rev Adv Mater Sci 30(3):225–42

Muzzarelli RAA (2011) Biomedical exploitation of chitin and chitosan via mechano-chemical disassembly, electrospinning, dissolution in imidazolium ionic liquids, and supercritical drying. Marine Drugs 9(9):1510–1533. https://doi.org/10.3390/md9091510

Orue A, Corcuera MA, Pena C, Eceiza A, Arbelaiz A (2014) Bionanocomposites based on thermoplastic starch and cellulose nanofibers. J Thermoplast Compos Mater 29:817–832. https://doi.org/10.1177/0892705714536424

Salaberria AM, Diaz RH, Labidi J, Fernandes SCM (2015a) Role of chitin nanocrystals and nanofibers on physical, mechanical and functional properties in thermoplastic starch films. Food Hydrocoll 46:93–102. https://doi.org/10.1016/j.foodhyd.2014.12.016

Salaberria AM, Labidi J, Fernandes SCM (2015b) Different routes to turn chitin into stunning nano-objects. Eur Polym J 68:503–515. https://doi.org/10.1016/j.eurpolymj.2015.03.005

Salehudin MH, Salleh E, Mamat SNH, Muhamad II (2014) Starch based active packaging film reinforced with empty fruit bunch (EFB) cellulose nanofiber. Procedia Chem 9:23–33. https://doi.org/10.1016/j.proche.2014.05.004

Tang X, Alavi S (2011) Recent advances in starch, polyvinyl alcohol based polymer blends, nanocomposites and their biodegradability. Carbohydr Polym 85:7–16. https://doi.org/10.1016/j.carbpol.2011.01.030

Tzoumaki MV, Moschakis T, Biliaderis CG (2011) Mixed aqueous chitin nanocrystal-whey protein dispersions: microstructure and rheological behaviour. Food Hydrocoll 25:935–942. https://doi.org/10.1016/j.foodhyd.2010.09.004

Visakh PM, Thomas S (2010) Preparation of bionanomaterials and their polymer nanocomposites from waste and biomass. Waste Biomass Valorization 1:121–134. https://doi.org/10.1007/s12649-010-9009-7

Yu J, Yang J, Liu B, Ma X (2009) Preparation and characterization of glycerol plasticized-pea starch/ZnO-carboxymethylcellulose sodium nanocomposites. Bioresour Technol 100:2832–2841. https://doi.org/10.1016/j.biortech.2008.12.045

Zia F, Zia KM, Zuber M, Kamal S, Aslam N (2016) Starch based polyurethanes: a critical review updating recent literature. Carbohydr Polym 134:784–798. https://doi.org/10.1016/j.carbpol.2015.08.034

# Chapter 2
# Processing Techniques

## 2.1 Starch Based Composites

### 2.1.1 Solution Casting

Solution casting is one of the most common processing techniques for the preparation of starch based bionanocomposite. Solution casting is an easy method for homogeneous dispersion of starch and reinforcement using water as the solvent.

Sreekumar et al. prepared nano $TiO_2$ filled starch/PVA composite by solution casting method (Sreekumar et al. 2012). Initially the suspension of nano $TiO_2$ in distilled water at various concentrations of nano $TiO_2$ (i.e. 0.5, 1, and 2 wt% with respect to the total polymer content) were prepared using ultrasonicator using the dispersing agent span 60. The suspension was ultrasonicated for 1 h and then added slowly to starch/PVA solution (1:1) containing 3.5 g of glycerol while stirring continuously. Composite films were obtained.

Almasi et al. prepared citric acid (CA) modified starch-carboxymethyl cellulose (CMC)-montmorillonite (MMT) bionanocomposite films by casting method (Almasi et al. 2010). Starch in distilled water was mixed with glycerol and CA at room temperature (25 °C) for 5 min. Suspension was agitated by magnetic stirrer. Starch CMC was solubilized in 75 mL of water at 75 °C for 10 min. On the other hand, montmorillonite (MMT) was dispersed in distilled water by sonication. The clay dispersion was added to the aqueous dispersion of starch. CMC and starch-MMT solution were mixed together and stirred. Dried films were then prepared by casting.

Chang et al. prepared glycerol plasticized potato starch (GPS) reinforced with chitin nanoparticle (CNP) by casting and evaporation method. At low loading levels, CNP were uniformly dispersed in the GPS matrix and had good interaction between the filler and matrix, which led to improvements in tensile strength, storage modulus, glass transition temperature, and water vapor barrier properties of the GPS/CNP composites. However, at higher loading (greater than 5 wt.%), aggregation of CNP had a negative effect on these properties (Chang et al. 2010).

© The Author(s), under exclusive licence to Springer Nature Switzerland AG 2019    9
M. S. Thomas et al., *Starch, Chitin and Chitosan Based Composites
and Nanocomposites*, Biobased Polymers,
https://doi.org/10.1007/978-3-030-03158-9_2

El Miri et al. prepared bionanocomposite films of carboxymethyl cellulose (CMC)/Starch (ST) polysaccharide matrix reinforced with cellulose nanocrystals (CNC) using solution casting method (El Miri et al. 2015). The CNC were extracted from sugarcane bagasse (SCB) via sulphuric acid hydrolysis process. The loading levels of CNC in CMC/ST-CNC bionanocomposite films were fixed at 0.5, 2.5 and 5.0 wt%. Enhancement in properties of composite film occurs due to strong interfacial adhesion generated from the hydrogen bonding between the functional groups that are present in the components CMC, starch and CNC. Figure 2.1 shows the visual observations of the bionanocomposite films.

Fabrication of pea starch (PS)/cellulose nanowhiskers from pea hull fiber (PHFNW) were done by solution casting and evaporation method. PHFNW dispersions were prepared and coded as PHFNW-t, where t was the hydrolysis time. For example, PHFNW-8 means the nanowhiskers were hydrolysed from PHF by sulphuric acid for 8 h. A series of pea hull fiber-derived nanowhiskers (PHFNW-t) was successfully extracted from pea hull fibers (PHF) by sulphuric acid hydrolysis with different hydrolysis times (t). The PHFNW-t was then blended with pea starch (PS) to prepare PS/PHFNW-t bionanocomposite films. Compared to the neat PS film and the PS/PHF (t = 0 h) film, the PS/PHFNW-t nanocomposite films showed higher ultraviolet absorption, transparency, tensile strength, elongation at break, and water-resistance (Chen et al. 2009).

Tape casting (De Moraes et al. 2013) is another method that is used for the preparation of thin films. The starch–glycerol–fiber suspensions were prepared as sketched in Fig. 2.2a. Tape casting technique allows the spreading of a suspension on large supports, controlling the thickness by an adjustable blade at the bottom of the spreading device. Drying of the films can be carried-out on the support itself, under controlled conditions. Cellulose fiber reinforced starch biocomposite films were prepared by this method. The obtained films were homogeneous, without bubbles or defects and were easy to handle, as shown qualitatively in Fig. 2.3.

**Fig. 2.1** Digital images of CMC/ST, CMC/ST-CNC films%. Reprinted with permission from El Miri et al. (2015). Copyright Elsevier

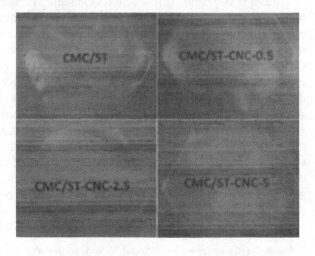

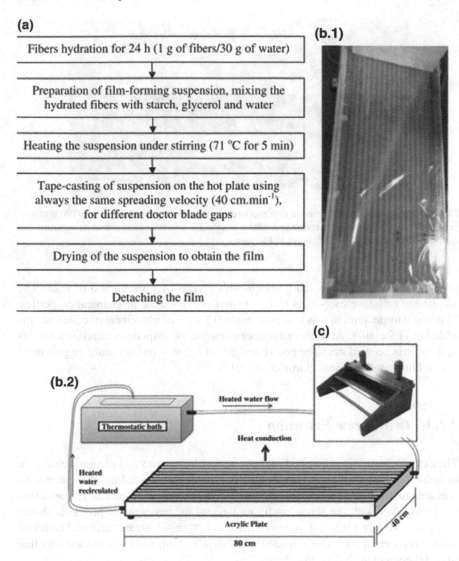

**Fig. 2.2** **a** Summary of the process used to prepare cellulose fibers–starch films by tape-casting, **b**.1 Picture of the support for the spread suspension in the discontinuous (manual) tape-casting device, **b**.2 Sketch of the acrylic plate, and **c** Picture of a doctor blade device (De Moraes et al. 2013). Reprinted with permission from De Moraes et al. (2013). Copyright Elsevier

## 2.1.2 Compression Moulding

Orue et al. prepared composites based on thermo plastic starch (TPS) and cellulose nanofibers by compression moulding method. Cellulose nanofibers were extracted by chemical treatments from sisal fibers. TPS composition with the best mechanical

**Fig. 2.3** Image of a large dimension film prepared by tape-casting, with 3 g starch/100 g of suspension, 0.20 g glycerol/g dry starch and 0.30 g fiber/g dry starch %). Reprinted with permission from De Moraes et al. (2013). Copyright Elsevier

property was observed for 60 g starch with glycerol to water ratio of 0.25. The addition of cellulose nanofibers to TPS matrix improved the mechanical properties. Maximum improvement was observed when 0.5 wt% of nanofibers of cellulose was added to TPS matrix. At higher nanofiber contents, the dispersion of reinforcements in the matrix seemed not to be good enough and agglomerations could be generated where film started to break (Orue et al. 2014).

### 2.1.3  Twin Screw Extrusion

The extruder barrel was equipped with two atmospheric vents and vacuum ventilation in order to remove the vaporized water from the material. Feeding of the materials was done manually due to the small amounts of the prepared pre-mixes. The extrusion set-up together with the screw configuration and the temperature profile is shown in Fig. 2.4. Before extrusion, premixes of starch, sorbitol, stearic acid and cellulose nanofibers were made. The extruded material was compression moulded into thin films (Hietala et al. 2013) (Fig. 2.5).

Nascimento et al. used the method of melt extrusion for the preparation of rice hull (RH), bleached rice hull (RCB) and cellulose nanofibrils from rice hull (RHNF) reinforced thermoplastic starch films. The mixture was fed to the extruder first to obtain pellets of gelatinized material. These pellets were then re-extruded for better homogenization. The second batch of pellets was fed to the extruder for film manufacturing by blowing.

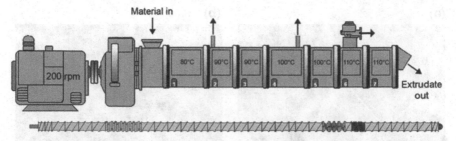

**Fig. 2.4** Extrusion set-up used in the compounding of TPS/CNF composites %). Reprinted with permission from Hietala et al. (2013). Copyright Elsevier

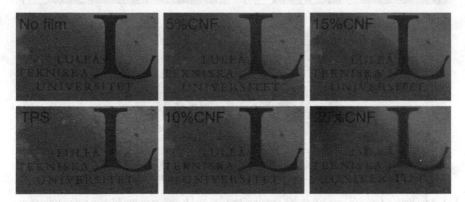

**Fig. 2.5** Visual appearance of TPS and TPS/CNF films %). Reprinted with permission from Hietala et al. (2013). Copyright Elsevier

### 2.1.4 Thermal Moulding

Lomelí Ramírez et al. (2011) prepared coir fiber reinforced cassava starch based composites by thermal moulding using glycerol as the plasticizer. With the help of a mixer, the coir fibers were well dispersed in the matrix starch. The quantity of coconut fibers used to prepare the composites were 5, 10, 15, 20, 25 and 30% (Souza et al. 2012). Figure 2.6a, b shows photographs of cassava starch–glycerol composites with 5 and 30% coir fibers as observed under a stereomicroscope.

### 2.1.5 Melt Blending

Castaño et al. prepared cellulosic reinforced composite by melt blending. *Araucaria araucana* (pehuen) cellulosic husk was employed as the reinforcement in pehuen thermo plastic starch (TPS). A comparative study of pehuen TPS/poly vinyl alcohol (PVA)/poly lactic acid (PLA) blends were also done. Incorporation of pehuen husk

**Fig. 2.6** Photographs of cassava starch composite **a** 5% and **b** 30% coconut fiber. **c** Panoramic view of the composite at higher magnification; TPS: thermoplastic starch; CF: coir fiber %). Reprinted with permission from Lomeli Ramirez et al. (2011). Copyright Elsevier

improved the thermal stability and mechanical properties of the studied composites considerably, mainly in TPS composites (Castaño et al. 2012).

## 2.2 Processing Techniques of Chitin and Nano Chitin Based Composite Films

The possibility of formation of continuous nanocrystal network and the final properties of nanocomposite material are governed by the processing method.

### 2.2.1 Solution Casting

Solution casting, also called wet process is a technique which involves the dispersion or solubilization of chitin in a solvent medium before pouring the solution onto a flat surface and drying. The film can be peeled off the flat surface after drying. This slow process of casting/evaporation are reported to give materials with the highest mechanical performance. This is attributed to the fact that during slow water evaporation rearrangement of nanoparticles is possible due to Brownian motion in the suspension or solution. Thus they get enough time to interact and connect to form a percolating network which is the basis of their reinforcing effect. The resulting structure is completely relaxed and direct contacts between nanocrystals or microfibrils are then created (Dufresne 2008).

## 2.2.2 *Extrusion*

Extrusion being a conventional polymer-processing technique there is growing interest in this technique for nanocomposite processing in industrial applications. The main challenges of this technique are the feeding of the nanomaterials into the extruder and obtaining well-dispersed nanocrystals in the polymer matrix since nanomaterials tend to aggregate while drying. Since chitin forms hydrogen bonds when dried, drying prior to the extrusion is not suitable. Liquid feeding of nanomaterials together with a suitable processing aid, such as plasticizer or water, is one way to avoid the drying process and to improve the dispersion of the nanomaterials in the matrix (Herrera et al. 2016). However, when water or other solvents are used, it is important to use a co-rotating twin-screw extruder where the steam can be effectively removed during the processing.

## 2.2.3 *Freeze-Drying Method*

The arrangement of nanoparticles in the suspension is first frozen during the freeze-drying/hot-processing method and during the subsequent hot-processing stage, the rearrangement of particles is strongly limited because of polymer melt viscosity and contacts can be made through a limited amount of polymer matrix. However, although the freeze-drying/hot pressing process limits the possibility of creation of hydrogen bonds, it is expected that, for high polysaccharide nanoparticles content, some bonds may still be created. Zheng et al. (2003) prepared nanocomposites with Soy Protein Isolate (SPI) using freeze drying method. SPI of desired weight and various content of chitin were mixed and stirred to obtain a homogeneous dispersion. The dispersion was freeze-dried and 30% glycerol was added. The resulting mixture was hot-pressed at 20 MPa for 10 min at 140 °C and then slowly cooled to room temperature.

## 2.3   Processing Techniques of Chitosan Based Composites

### 2.3.1   Physical Blends

Polymeric composites can be prepared by the physical mixing of two or more polymers. But it is very difficult to prepare a homogeneous system of chitosan and polyester because of lack of co-solvents that can accommodate both polymers (Levengood and Zhang 2014). Mechanical stirring was utilized by Jiang et al. (2006) to fabricate a homogenous suspension of milled chitosan microparticles in a solution of PLGA in methylene chloride. Subsequently, composite chitosan-PLGA microspheres were formed via solvent evaporation and microspheres were fused to yield scaffolds (Jiang et al. 2006). Melt blending techniques were also adopted for the preparation of chitosan—polyester composites where heating and compression of polymer particles results in the formation of a continuous polymer network (Levengood and Zhang 2014). Combined with leaching of salt particles, melt blending yields porous scaffolds and in both cases, mechanical strength of the resultant scaffolds exceeded that of pure chitosan scaffolds (Levengood and Zhang 2014).

### 2.3.2   Polyelectrolyte Complexes

The electrostatic interactions between charged functional groups chitosan and anionic macromolecules results in the formation of polyelectrolyte complex (PEC) networks. Mechanical stirring was utilized by Jiang et al. (2006) to fabricate a homogenous suspension of milled chitosan microparticles in a solution of PLGA in methylene chloride. Subsequently, composite chitosan-PLGA microspheres were formed via solvent evaporation and microspheres were fused to yield scaffolds (Jiang et al. 2006). Melt blending techniques were also adopted for the preparation of chitosan—polyester composites where heating and compression of polymer particles results in the formation of a continuous polymer network (Levengood and Zhang 2014).

## References

Almasi H, Ghanbarzadeh B, Entezami AA (2010) Physicochemical properties of starch–CMC—nanoclay biodegradable films. Int J Biol Macromol 46(1):1–5.

Castaño J, Rodríguez-Llamazares S, Carrasco C, Bouza R (2012) Physical, chemical and mechanical properties of pehuen cellulosic husk and its pehuen-starch based composites. Carbohydr Polym 90:1550–1556. https://doi.org/10.1016/j.carbpol.2012.07.029

Chang PR, Jian R, Yu J, Ma X (2010) Starch-based composites reinforced with novel chitin nanoparticles. Carbohydr Polym 80:421–426. https://doi.org/10.1016/j.carbpol.2009.11.041

Chen Y, Liu C, Chang PR, Cao X, Anderson DP (2009) Bionanocomposites based on pea starch and cellulose nanowhiskers hydrolyzed from pea hull fibre: effect of hydrolysis time. Carbohydr Polym 76:607–615. https://doi.org/10.1016/j.carbpol.2008.11.030

De Moraes JO, Scheibe AS, Sereno A, Laurindo JB (2013) Scale-up of the production of cassava starch based films using tape-casting. J Food Eng 119:800–808. https://doi.org/10.1016/j.jfoodeng.2013.07.009

Dufresne A (2008) Polysaccharide nano crystal reinforced nanocomposites. Can J Chem 86:484–494. https://doi.org/10.1139/v07-152

El Miri N, Abdelouahdi K, Barakat A, Zahouily M, Fihri A, Solhy A, El Achaby M (2015) Bio-nanocomposite films reinforced with cellulose nanocrystals: rheology of film-forming solutions, transparency, water vapor barrier and tensile properties of film. Carbohydr Polym 20(129):156–167. https://doi.org/10.1016/j.carbpol.2015.04.051

Herrera N, Salaberria AM, Mathew AP, Oksman K (2016) Plasticized polylactic acid nanocomposite films with cellulose and chitin nanocrystals prepared using extrusion and compression molding with two cooling rates: Effects on mechanical, thermal and optical properties. Compos Part A Appl Sci Manuf 83:89–97. https://doi.org/10.1016/j.compositesa.2015.05.024

Hietala M, Mathew AP, Oksman K (2013) Bionanocomposites of thermoplastic starch and cellulose nanofibers manufactured using twin-screw extrusion. Eur Polym J 49:950–956. https://doi.org/10.1016/j.eurpolymj.2012.10.016

Jiang T, Abdel-Fattah WI, Laurencin CT (2006) In vitro evaluation of chitosan/poly (lactic acid—glycolic acid) sintered microsphere scaffolds for bone tissue engineering. Biomater 27(28):4894–903. https://doi.org/10.1016/j.biomaterials.2006.05.025

Levengood SL, Zhang M (2014) Chitosan-based scaffolds for bone tissue engineering. J Mater Chem B 2(21):316. https://doi.org/10.1039/C4TB00027G

Lomelí Ramírez MG, Satyanarayana KG, Iwakiri S, de Muniz GB, Tanobe V, Flores-Sahagun TS (2011) Study of the properties of biocomposites. Part I. Cassava starch-green coir fibers from Brazil. Carbohydr Polym 86:1712–1722. https://doi.org/10.1016/j.carbpol.2011.07.002

Orue A, Corcuera MA, Pena C, Eceiza A, Arbelaiz A (2014) Bionanocomposites based on thermoplastic starch and cellulose nanofibers. J Thermoplast Compos Mater 29:817–832. https://doi.org/10.1177/0892705714536424

Souza AC, Benze R, Ferrão ES, Ditchfield C, Coelho ACV, Tadini CC (2012) Cassava starch biodegradable films: influence of glycerol and clay nanoparticles content on tensile and barrier properties and glass transition temperature. LWT Food Sci Technol 46:110–117. https://doi.org/10.1016/j.lwt.2011.10.018

Sreekumar P, Al-Harthi M, De SK (2012) Reinforcement of starch/polyvinyl alcohol blend using nano-titanium dioxide. J Compos Mater 46:3181–3187. https://doi.org/10.1177/0021998312436998

Zheng H, Tan Z, Zhan YR, Huang J (2003) Morphology and properties of soy protein plastics modified with chitin. J Appl Polym Sci 90:3676–3682. https://doi.org/10.1002/app.12997

# Chapter 3
# Properties of Composites

## 3.1 Thermal Properties of Composites

### 3.1.1 Thermal Properties of Starch Based Composites

Thermal stability of composites depends on the type of reinforcement added to the starch. Study of thermal stability of glycerol-plasticized pea starch/graphene oxide (PS/GO-n) biocomposite films with different loading levels of graphene oxide revealed that an improvement in the stability of biocomposite films can be seen with an increase of the fillers loading (Li et al. 2011). Thermogravimetric (TG) and derivative thermogravimetric (DTG) curves of the PS film and PS/GO-n biocomposites are shown in Fig. 3.1a, b, respectively.

Differential Scanning Calorimetry (DSC) is widely accepted as the most suitable method for the evaluation of starch gelatinization and other thermal behaviour. DSC thermograms facilitate the analysis of starch transition temperatures as well as transition enthalpies. The enthalpy ($\Delta H$) of a transition has been interpreted as corresponding to the amount of crystal order in starch suspensions. The reported DSC results are not always consistent and are sometimes controversial, not only due to the complexity of the thermal behaviors of starches, but also due to the differing measurement conditions used. The key factors that affect the measured results are sample preparation, the type of pan used and the measurement conditions. More recently, a new technique using high-pressure stainless-steel pans has been developed, which allows the study of the phase transitions of starch with higher water content (9–75%) and at higher temperatures (up to 360 °C). In DSC experiments, the glass transition temperature ($T_g$) is generally taken as the inflection point of the specific heat increment at the glass-rubber transition. Incorporation of filler in thermoplastic starch affects the glass transition temperature (Gironès et al. 2012). Homogeneous dispersion of flax cellulose nanocrystals within the pea starch matrix and strong interfacial adherence between matrix and fillers led to an increase of glass

M. S. Thomas et al., *Starch, Chitin and Chitosan Based Composites and Nanocomposites*, Biobased Polymers, https://doi.org/10.1007/978-3-030-03158-9_3

19

**Fig. 3.1** TG (**a**) and DTG
(**b**) curves of PS and
PS/GO-n biocomposite
films). Reprinted with
permission from Li et al.
(2011). Copyright Elsevier

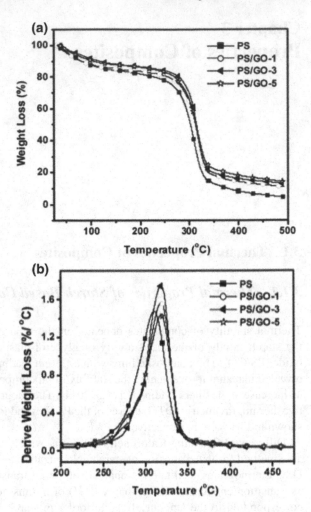

transition temperature ascribed to the starch molecular chains in the starch-rich phase
(Cao et al. 2008).

## 3.1.2  *Thermal Properties of Chitin Based Composites*

Thermal properties of materials are of importance for processing issues and practical
applications. The thermal behavior of chitin and chitin/ZnO composite was determined by thermogravimetry and differential thermal analysis by Wysokowski et al.
(2013a, b). The main thermal degradation of the biopolymer β chitin started at 200 °C
and ended at 550 °C and two steps were strongly exothermic. First exothermic effect

is associated with the thermal depolymerization of chitin and the formation of volatile low molecular products, which burn in the oxidative atmosphere of air. The second one is due to the oxidative thermal degradation (burning) of the formed char. Thermogravimetric and differential thermal analyses proved that thermal decomposition of chitin/ZnO composite started at the same temperature as in the case of β-chitin, but exothermic polymer degradation occurred at the range of 200–650 °C. This clearly demonstrated that hydrothermal deposition of ZnO on chitin led to development of materials with higher thermal stability.

The thermal stability of the final thermoplastic nanocomposite films was dependent of the kind of chitin nano-object incorporated. The thermal properties of chitin crystals were found to be higher than the thermal stability of the cellulose crystals (Herrera et al. 2016). Chitin nanocrystal and chitin nanofiber were found to exhibit different thermal reinforcing effect on the polymer matrix. Incorporation of CHNC into the thermoplastic starch matrices had a negative effect on the thermal stability of the final S/CHNC nanocomposite films (Salaberria et al. 2015a, b). The maximum degradation temperature decreased up to 15 °C—from 285 °C for S to 270 °C for S/CHNC20. The decrease in flexibility of amylopectin chains in the presence of crystalline CHNC may be the main reason for the decrease of the thermal stability of S/CHNC nanocomposite films. Contrarily, S/CHNF nanocomposite films demonstrated a positive effect on the thermal stability with the incorporation of CHNF (up to 5 °C). Certainly, the superior thermal stability of CHNF compared to CHNC was an important reason for the enhanced thermal stability of the obtained S/CHNF nanocomposite films (Salaberria et al. 2014).The addition of chitin nano whiskers (CNW) into starch matrix was also found to decrease the thermal stability of the nanocomposite film from 322 °C in the case of pure starch film to 309 °C in the case of nanocomposite film (Qin et al. 2016). Figure 3.2 shows the thermogravimetric (TGA) and derivative (dTGA) thermograms of S, CHNC, CHNF and S/CHNC and S/CHNF thermoplastic starch nanocomposite films.

Contrary to the above result, addition of chitin nanoparticle into gelatin films was found to increase the thermal stability of gelatin films from 160 to 169 °C (Sahraee et al. 2017). Therefore, it can be concluded that when chitin nanoparticles disperse and interact properly with gelatin matrix, thermal stability of the film is promoted. Chitin nanowhisker was found to be immiscible in polyvinylalcohol (PVA) matrix and this was confirmed by differential scanning calorimeter (DSC) analysis of chitin reinforced polyvinylalcohol matrix (Kadokawa et al. 2011; Uddin et al. 2012). The DSC profiles of composite films of PVA reinforced with chitin nanowhisker exhibited endothermic peaks due to a melting point of PVA. These results suggested PVA in the composites had some crystallinity, indicating immiscibility toward chitin. The melting points of PVA were shifted to lower temperatures accompanied with broadening of peaks when chitin in the composites increased. These data indicated that the crystallinities of PVA were decreased in the composites with increasing contents of chitin. This phenomenon substantiated the interaction between PVA and CNW, wherein PVA chains adhered with the CNW surface, which were likely to inhibit crystal growth by restricting the mobility of surrounding PVA chains.

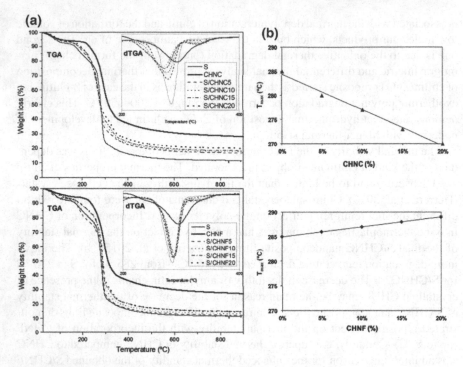

**Fig. 3.2  a** Thermogravimetric (TGA) and derivative (dTGA) thermograms of S, CHNC, CHNF and S/CHNC and S/CHNF thermoplastic starch nanocomposite films; and **b** effect of nanocrystals (top) and nanofibers (bottom) percentage on maximum degradation temperature of S/CHNC and S/CHNF thermoplastic starch nanocomposite film). Reprinted with permission from Salaberria et al. (2015a, b). Copyright Elsevier

Wang et al. (2012) evaluated the crystallizing and melting behaviour of poly(3-hydroxybutyrate-co-3-hydroxyvalerate) (PHBV) added with chitin nanocrystals using DSC. Crystallization temperatures were seen to drop with the increase of original chitin nanocrystal content. This is due to the hydrogen bonds between the PHBV carbonyls and the chitin hydroxyls which confine the diffusion and migration of polymer chains. Crystallization temperature (Tc) of the blend decreased to 65 °C with the addition of 1 wt% modified chitin nanocrystals. This shows that the hydrogen bonds between the PHBV carbonyls and the chitin hydroxyls did not decrease sufficiently because the surface modifications of chitin nanocrystals were limited.

Qin et al. (2016) studied the effects of CNW on the thermal properties of maize starch nanocomposite films with DSC analysis. Compared to the pure maize starch films, a higher on set temperature ($T_o$) and peak temperature ($T_p$) of the maize starch films with CNW were observed. Enthalpy of the composite films was observed to increase with the addition of CNW. These results may be due to the interactions between CNW and the chain segments of the maize starch, which might have led to an increase in the crystallinity of the resulting films. Therefore, it could be deduced that higher the gelatinization enthalpy (H), higher would be the compatibility of maize starch and CNW.

### 3.1.3 Thermal Properties of Chitosan Based Composites

Neto et al. reported that, weight loss of chitosan took place in two stages. The first one started at 60 °C. The second stage started at 240 °C and reached a maximum at 380 °C. The first stage is assigned to the loss of water. The second one corresponds to the decomposition (thermal and oxidative) of chitosan, vaporization and elimination of volatile products (Neto et al. 2005).

TGA analysis of $Fe_3O_4$/MWNT/Chitosan nanocomposites indicated that nanocomposites were degraded in two stages; the major thermal degradation appearing in the 120–400 °C range was caused by the depolymerization of chitosan chains through deacetylation and cleavage of glycosidic linkages via dehydration and deamination (Marroquin et al. 2013). The decarboxylation of carbon nanotube (CNT) defective sites has also been reported in this temperature range. The second degradation stage (400–700 °C) consisted of the thermal destruction of the pyranose ring to produce formic, acetic, and butyric acids as well as a series of lower fatty acids. Further degradation occurred for disordered carbon derived from thermal oxidation of CNTs and chitosan. There was another degradation stage at 750–800 °C for $Fe_3O_4$ composites which would correspond to the reduction of $Fe_3O_4$ by reaction with residual carbon (Marroquin et al. 2013).

TGA revealed that the thermal stability of chitosan films improved remarkably by the addition of nano-MgO (De Silva et al. 2017). The enhancements could be attributed to the high thermal stability of MgO (thermal decomposition starts around 2800 °C) as well as the uniform distribution of MgO in the chitosan matrix. In addition MgO can act as a barrier to hinder the diffusion of thermally degraded products of chitosan, which results in an effective delay in mass transport.

Three weight loss regions were observed by Archana et al. in the blends of chitosan, poly(N-vinylpyrrolidone) (PVP) and titanium dioxide ($TiO_2$) (Archana et al. 2013). For nanocomposite, weight loss at 65–140 °C with a weight loss of 12% was due to the loss of water. The second stage started at 170–370 °C with a weight loss of 47.5% due to the decomposition (thermal and oxidative) of chitosan and PVP. The third stage started at 400–500 °C with a weight loss of 11% due to grafting of nano $TiO_2$ into chitosan PVP matrix.

## 3.2  Mechanical Properties

### 3.2.1  Mechanical Properties of Starch Based Composites

Mechanical properties of TPS films depend on the ratio of water to glycerol. Tensile properties decrease with increasing ratio of glycerol to water. Glycerol can create interactions with starch chains and weaken the hydrogen bonding interaction between starch molecules facilitating the slippage among starch molecules and consequently producing a material with lower stiffness. TPS film with glycerol to water ratio of 0.25 showed the best mechanical properties as per the studies conducted by Orue et al. (2014).

The mechanical properties of composites may be associated with mainly geometrical ratio of filler, processing methods, compatibility of filler with matrix and filer to filler interaction (Pandey et al. 2009). Orue et al. found that the addition of cellulose nanofibers (extracted by chemical treatment from sisal fibers) to TPS matrix improved the mechanical properties. Maximum improvement was observed when 0.5 wt% of cellulose nanofibers was added to TPS matrix (Orue et al. 2014).

Harfiz et al. (2014) studied the mechanical property of starch films reinforced with cellulose nanofiber extracted from empty fruit bunch. The polymer strength increased with the addition of cellulose nanofiber. However, the optimum condition for effective reinforcement was detected at 2% of cellulose nanofiber incorporation. Further addition of cellulose nanofiber did not give significant improvements on film strength, elongation at break and Young's modulus. The nature of asymmetric and non-homogeneous natural cellulose created unequal dispersion in the film matrices lowering its mechanical strength.

Nascimento et al. studied the effect of rice hull (RH), bleached rice hull (RHB) and cellulose nanofibrils extracted from rice hull (RHNF) on the mechanical properties of starch films. Addition of RH and RHB resulted in a decrease in tensile strength, which might be associated with the non-homogeneity of the film matrix. Addition of RHNF resulted in an increase in tensile strength and elongation, and a decrease in Young's modulus (Nascimento et al. 2016).

Ramírez et al. (2011) studied the mechanical property of green coconut fibers reinforced starch based composites, where the matrix and composites were given thermal treatment.. It has been found that the tensile properties of cassava starch improved with both the incorporation of fibers and thermal treatment. Both the untreated and treated matrices and their composites were characterized for their tensile properties.

Figure 3.3 shows typical force–displacement curves obtained in tensile tests of TPS based materials, viz. TPS matrix with and without treatment (0% fiber), and two of their composites containing 15 and 30% coir fibers. The matrix (with or without treatment) showed typical curves of a ductile material with large deformation and low strength. The treated matrix showed higher breaking load and lower displacement compared to the untreated matrix. The composites subjected to thermal treatment at 60 °C showed typical curves of a rigid, strong but fragile material. Elongation decreased with increasing fiber content for both types of composites (treated and

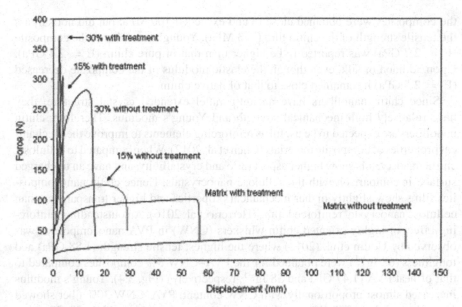

**Fig. 3.3** Force–displacement curves of TPS matrix with and without treatment as well as those of their composites containing 15 and 30% coir fiber. Reprinted with permission from Ramírez et al. (2011). Copyright Elsevier

untreated) (Ramírez et al. 2011). The study of mechanical property of cassava starch based composites reinforced with cinnamon oil showed that there occurs a decrease in tensile strength and elongation at break of films as a result of incorporation of cinnamon essential oil, indicating a loss of macromolecular mobility (Souza et al. 2013).

Tensile strength and Young's modulus of starch based nanocomposites reinforced with flax cellulose nanocrystals (FCNs) increased, respectively, from 3.9 to 11.9 MPa and from 31.9 to 498.2 MPa with an increase of FCNs content from 0 to 30 wt%. This has been explained as the reinforcement effect from the homogeneously dispersed high-performance FCNs as fillers in the starch matrix and the strong hydrogen bonding interaction between cellulose nanocrystals and starch molecules. The cellulose nanocrystals from flax fiber showed an effect similar to that of ramie-based on the mechanical properties in the starch based nanocomposites (Cao et al. 2008).

## 3.2.2 Mechanical Properties of Chitin Based Composites

The mechanical properties of chitin-silk biocomposite solutions of squid pen β-chitin and *B. mori* cocoon silk co-dissolved in hexa fluoro isopropanol (HFIP) was analysed by Jin et al. and he could confirm the occurrence of a strong interaction between chitin nanofiber phase and silk matrix (Jin et al. 2013). Tensile strength of the composites were obtained close to and exceeded 100 MPa, but did not surpass the tensile strength of the chitin film (113 MPa). Young's modulus of the composite (E = 2.7 GPa) was reported to be higher than that of pure chitin (E = 2.4 GPa). Upon addition of silk, even though the elastic modulus of the composite decreased (E = 2.5 GPa) it remained close to that of native chitin.

Since chitin nanofibers have an antiparallel extended crystal structure, they have relatively high mechanical strength and Young's modulus. Therefore, chitin nanofibers are expected to be useful as reinforcing elements to improve the mechanical properties of composite materials (Chen et al. 2012). When compared to cellulose, chitin nanocrystals show higher aspect ratio and crystallinity, and have an uncharged surface in comparison with the cellulose nanocrystals. Hence chitin nanocomposites films show slightly higher mechanical properties and higher transparency than cellulose nanocrystal reinforced films (Herrera et al. 2016). An outstanding reinforcing effect of highly oriented chitin whiskers (CNW) in PVA nanocomposites was observed by Uddin et al. (2012) where the highest tensile strength (1.88 GPa) and toughness (68 Jg$^{-1}$) were obtained for the PVA–CNW 5% composites compared to that of neat PVA (1.47 GPa and 58 Jg$^{-1}$, respectively) (Fig. 3.4). Young's modulus increased almost proportionally with CNW content. PVA–CNW 30% fiber showed Young's modulus of 50 GPa, much higher than the same of neat PVA, 28 GPa.

Tensile property of the electrospun nanocomposite PVA/$\alpha$-chitin whisker(CW) fiber mats and its corresponding solution cast nanocomposite PVA/$\alpha$-chitin whisker films were studied by Junkasem et al. (2006). It was observed that the tensile strength

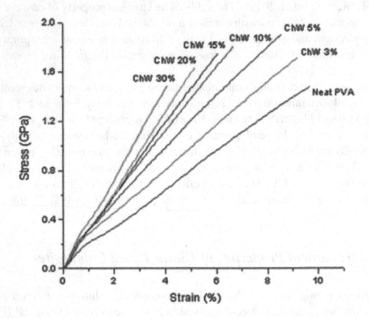

**Fig. 3.4** Representative stress–strain curves of neat PVA and PVA–CNW fibers. Reprinted with permission from Uddin et al. (2012). Copyright Elsevier

of the electro spun nanocomposite fiber mats increased from that of the neat PVA fiber mat (i.e., 4.3 ± 0.7 MPa) with initial addition of chitin whiskers to reach a maximum value (i.e., 5.7 ± 0.6 MPa) at the CW/PVA ratio of about 5.1% and decreased with further increase in the chitin whisker content. Similar to the nanocomposite fiber mats, incorporation of chitin whiskers increased the modulus of nanocomposite films from that of the neat PVA film. Specifically, the Young's modulus increased monotonically from that of the neat PVA fiber mat (i.e., 48.4 ± 11.6 MPa) to reach a maximum value (1500 ± 450 MPa) at the maximum CW/PVA ratio of about 25.4%.

Mechanical properties of chitin nanofibril reinforced carrageenan films determined by the tensile test (Shankar et al. 2015) showed an increase from 30.2 ± 1.8 to 44.7 ± 3.6 MPa with an increase in CNF concentration from 0 to 5 wt%, followed by a decrease with further addition of CNF. On the other hand, tiffness of the films determined by its Young's modulus increased linearly with increase in the CNF content. Increase in tensile strength and Young's modulus of the composite films blended with CNF can be attributed to the reinforcement effect of homogeneously dispersed high-strength CNF in the carrageenan polymer matrix. The reinforcing effect of CNF is mainly due to their high stiffness and strength and the increase in the mechanical properties of polymer/CNF nanocomposite films resulted from the formation of a percolating network based on hydrogen bonding forces. The decrease in tensile strength with high content of CNF is due to the agglomeration of CNF or non-homogeneous dispersion of CNF at high concentrations caused by the self-networking of nanowhiskers.

According to a study carried out by Sahraee et al., increasing the concentration of nano chitin gel (N-chitin) up to 5%, the ultimate tensile strength of the films increased from 65.19 MPa for neat gelatin films to 119.08 MPa by, but increase of N-chitin concentration up to 10% decreased their ultimate tensile strength (Sahraee et al. 2017). Similarly the Young's modulus also enhanced for the films containing up to 5% N-chitin but decreased by further increase in N-chitin concentration. The increased tensile result shows the enhanced stiffness of the film as a result of N-chitin reinforcement effect. The N-chitin reinforcement effect in the matrix be attributed to different mechanisms such as higher chitin nanoparticles stiffness and density in comparison to protein matrix, filling of empty spaces in amorphous regions, creating a strong network by enhanced hydrogen bonds and increase of crystallinity in matrix. Similarly, Lu et al. (2004) used N-chitin in soy protein polymer and found out that by increasing filler concentration up to 20% of dry matter improved the polymer mechanical properties.

The tensile strength of starch films that contained CNWs were significantly ($P < 0.05$) higher than that of pure maize starch films (Qin et al. 2016). Furthermore, tensile strength of the composite films increased to the maximum with an increase in filler content up to 1%; however the value of tensile strength decreased slightly with more than 1% of CNWs. Similarly, Chang et al. (2010) also observed that the tensile strength of potato starch/chitin nanoparticle composite films increased with an increase in the filler content up to 6% and then decreased beyond 6% of chitin nanoparticles. The enhanced mechanical performance of the resulting starch-based nanocomposites could be due to the formation of a rigid network of the CNW (due to the interaction among the chitin nano size fillers by intra- and intermolecular hydrogen bonds) and/or the mutual entanglement between the chitin nano size fillers and the starch matrix. A similar increase in tensile strength and modulus was observed for chitin nanocrystal and chitin nanofiber reinforced starch matrix (for instance, for S/CHNC samples with 5 and 20% of CHNC, Young's modulus of S/CHNC nanocomposite films increased from 25 to 75 MPa, while the tensile strength increased from 2 to 3 MPa) (Fig. 3.5) (Salaberria et al. 2015a, b). The OH groups of starch exerted an important role with respect to the interaction among OH and residual NH$_2$ groups of CHNC, possibly due to the formation of intermolecular hydrogen bonds, which ensured the integrity of the continuous structure inside the matrix. In the case of S/CHNF samples, Young's modulus and tensile strength jumped from 25 to 425 and from 5 to 11 MPa respectively, in the thermoplastic starch nanocomposite films prepared with 5 and 20 wt% of CHNF, respectively. In this case, the increase in mechanical properties can be attributed to the percolation effect, since the formation of intermolecular hydrogen bonds were masked by the long and highly entangled CHNF morphology. However, the nanocomposite materials processed by melt-mixing approach showed lower mechanical performance (Salaberria et al. 2014). Moreover, these data confirmed that the mechanical properties of thermoplastic starch nanocomposite films improved in a morphological-dependent manner.

Addition of 0.15% w/w chitin whisker into alginate fibers was found to increase the tenacity of the nanocomposite fibers to reach a maximum value. A similar trend was observed on the elongation at break of these fibers, with the maximum value being observed at a slightly different whisker content from that of the tenacity (i.e., at 0.10% w/w) (Watthanaphanit et al. 2008). Rubentheren et al. (2015) observed that the addition of chitin whiskers into chitosan matrix was found to sharply increase the tensile strength from 22.02 to 52.23 MPa and significantly reduce the elongation at break from 50.3 to 21.32%. Addition of chitin whiskers increased the Young's modulus more than twice than that of the chitosan matrix. Chen et al. (2012) found that the incorporation of chitin nanofiber effectively improved the mechanical properties of polymethylmethacrylate (PMMA). When the fiber content increased from 40 to 70

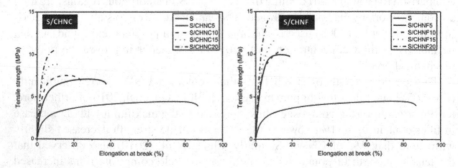

**Fig. 3.5** Stress–strain diagram of thermoplastic starch matrix (S) and thermoplastic starch-based nano-biocomposites (S/CHNC and S/CHNF) containing 5, 10, 15 and 20 wt% of chitin nano-size fillers %). Reprinted with permission from Salaberria et al. (2014). Copyright Elsevier

wt%, the tensile strength and Young's modulus of the corresponding nanocomposites enhanced from 77.5 to 95.8 MPa and from 3.14 to 4.46GPa, while that of native PMMA was only 41.8 MPa and 1.5 GPa, respectively. This significant reinforcement was due to the entanglement and confinement effect of ultra-long chitin nanofibers in PMMA matrix.

In a study carried out by Herrera et al., the yield strength and the Young's modulus of triethyl citrate (TEC) plasticized polylactic acid (PLA) films was observed to significantly increased with the addition of 1 wt% chitin nanocrystals (CNC). These properties were improved by 478% (from 3.7 to 21.4 MPa) and by 300% (from 0.3 to 1.2 GPa) respectively, with the addition of CNC. The toughness and elongation to break of the nanocomposites decreased in comparison with that of the plasticized PLA, probably due to the higher degree of the crystallinity of CNC (Herrera et al. 2016). Results also showed that the cooling rates significantly affected crystallinity of the materials and, consequently, the transparency of the films and elongation to break. Fast cooled films showed to be the most transparent materials possessing the highest elongation to break with values around 300%.

### 3.2.3 Mechanical Properties of Chitosan Based Composites

Regardless of numerous advantages and exceptional properties of chitosan, its mechanical properties are not good enough to satisfy a wide range of applications. The formation of organic–inorganic hybrids through incorporation of fillers is an effective approach for improving physical and mechanical properties of chitosan. Hence several reinforcing agents such as hydroxyapatite (HA), clay, carbon nanotubes, graphene oxide, etc. have been used.

According to Hu et al. (2004), the chitosan/hydroxyapatite (CS/HA) composites prepared by in situ hybridization exhibited mechanical properties 2–3 times stronger than that of PMMA and bone cement. Meanwhile, the bending strength and elastic modulus of CS/HA was only 34 and 17% when compared to cortical bone pin made from human femora. Incorporation of HA into CS matrix via blending method resulted in the decrease of mechanical properties of CS/HA material due to the weaker interfacial bonding between HA filler and CS matrix. However, there was no decrease of mechanical properties of CS/HA prepared by in situ hybridization, with values similar to that of pure CS.

The tensile strength and tensile modulus were significantly improved for the PLA/ENR/CS composites with the addition of CS into the matrix, while the elongation at break decreased (Zakaria et al. 2013). The tensile strength increased up to 5 wt% CS loading for both PLA/CS and PLA/ENR/CS and thereafter decreased while Young's modulus increased up to 10 wt%. However, when the CS content was increased to 15 wt%, the tensile strength and tensile modulus slightly decreased. These improvements were attributed to the good dispersion of CS at the optimum filler levels and attractive interaction between the composite components.

Marroquin et al. studied the tensile behavior of $Fe_3O_4$ /MWNT/Chitosan nanocomposite prepared by a simple solution evaporation method. A significant synergistic effect of $Fe_3O_4$ and MWNT enhanced the mechanical properties of the nanocomposites. They found that the tensile strength and modulus of MWNT/chitosan nanocomposites improved by adding $Fe_3O_4$. The best tensile strength was shown by a 5% (wt) loading. DMA analysis showed that due to the anti-plasticizing effect of $Fe_3O_4$, the composites exhibited higher crystallinity as well as restricted mobility of chitosan chain segments, thus further improving the mechanical properties of the nanocomposite films (Marroquin et al. 2013).

Transparent, biocompatible and biodegradable chitosan (CS) nanocomposite films reinforced with nanofibrillar cellulose (NFC) were prepared by means of solution casting method by Wu et al. (2014). The mechanical properties of NFC–CS nanocomposites were dependent on the content of NFC. Under dry conditions, the yield strength and Young's modulus of chitosan nanocomposite films increased with increasing NFC content because of the reinforcement of NFC, while the elongation at break decreased. The improvements in ultimate strength and Young's modulus for wet samples were larger than those for dry samples.

De Silva et al. (2017) fabricated nano-MgO reinforced chitosan nanocomposites. The tensile strength of chitosan films significantly improved with the addition of MgO. For 5% MgO the tensile strength increased by 86 and 26% with the addition of 10 (w/w %) MgO, to that of pure chitosan films. This significant improvement could be attributed to the stronger interfacial interaction between the hydroxyl and amine groups of chitosan with MgO via hydrogen bonding. Superior results observed for composites with 5 (w/w %) loading in comparison to that of 10 (w/w %) might be due to the differences in degree of dispersion of MgO within the chitosan matrix. When there are agglomerated nanoparticles, the stress concentration is considerably high around the aggregates and the material would be easily ruptured under an applied force.

Kumar et al. (2012) prepared flexible and microporous chitosan hydrogel/nano ZnO composite bandages for wound dressing. Tensile strength of the composites were measured, and chitosan control exhibited a value of 0.1 MPa, which was adequate for a wound dressing material. Owing to the interaction between ZnO and chitosan, the tensile strength of composite material increased slightly up to 0.15 MPa. The obtained results showed that the CZBs had sufficient strength to bear the force applied on them. Control and CZBs showed elongation in the range of 40–60% at break points.

## 3.3 Morphological Analysis

### 3.3.1 Morphological Analysis of Starch Based Composites

Scanning electron microcopy (SEM) was used to evaluate films homogeneity, layer structure, pores and cracks, surface smoothness and thickness. TPS films presented homogeneous and smooth surfaces, without visible pores and cracks.

Morphological study of biodegradable films based on thermoplastic native corn starch containing chitosan and chitin showed that the films had homogeneous and smooth surfaces, without pores and cracks and no glycerol migration. Chitosan and chitin incorporation to TPS matrix induced some structural modifications due to the interactions between starch hydroxyl and chitosan/chitin amino groups (Lopez et al. 2014).

Surface analysis of TPS films reinforced with rice hull fiber showed that the control film, RH-film and RHB-film had rough surfaces, without any fibers of agglomeration. The RHNF-film demonstrated a more homogeneous and smooth surface than the other samples. The nanofibers were well dispersed and covered by the matrix, suggesting that the introduction of cellulose nanofibers into the starch-glycerol films resulted in improved adhesion between the nanofibers and the polymer matrix. The surfaces of the extruded films were examined by SEM (Fig. 3.6) (Nascimento et al. 2016).

### 3.3.2 Morphological Analysis of Chitin Based Composites

A new method of synthesis of chitin–inorganic hybrid materials is biomimetic synthesis wherein chitin acts as a template for hydrothermal synthesis of advanced materials like ZnO, $GeO_2$, silica etc. SEM and TEM can be used to observe the reaction pathways. Wysokowski et al. (2015a, b) used SEM images to understand the hydrothermal formation of germanium oxide from a germanium tetraethoxide

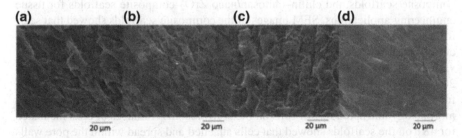

**Fig. 3.6** SEM images of the control film (**a**), RH-film (**b**), RHB-film (**c**) and RHNF-films (**d**). Reprinted with permission from Nascimento et al. (2016). Copyright Elsevier

(TEOG) precursor on the surface of α-chitin from the *Aplysina cauliformis* sponge leading to the formation of chitin–$GeO_2$ composites. SEM images show that $GeO_2$ nanoparticles are tightly bonded to the nano porous chitin surface which could not be removed even after 1 h of ultrasound treatment. Moreover, images of mechanically fractured fragments of the composite obtained from SEM measurements show that nanofibrils of chitin about 17 nm in diameter represent the nucleation sites for growth and formation of $GeO_2$ nanocrystals that are up to 300 nm in size. SEM micrographs could also be used to study the hydrothermal formation of ZnO using the β-chitin and it was observed that the hydrothermal formation of ZnO led to the formation of uniform nano rods with a sharp top and is similar to nano rods which grow without any chitin template. While ZnO nanocrystals originate from the surface of chitin and are tightly bound to this substrate by nanofibrils, they are structurally identical to those obtained without the organic matrix (Wysokowski et al. 2013a, b). Chitin-based scaffolds isolated from the skeleton of marine demosponge *Aplysina aerophoba* were used as a template for the in vitro formation of iron oxide from a saturated iron (III) chloride solution under hydrothermal conditions and the SEM micrograph was used to confirm that the method allowed an efficient growth of inorganic nanostructures within the chitinous scaffolds (Wysokowski et al. 2015a, b).

Wysokowski et al. (2013a, b) also prepared chitin–silica composites by in vitro silicification of two-dimensional *Ianthella basta* demosponge chitinous scaffolds under modified Stöber conditions and they found that spherical shaped $SiO_2$ particles were well-dispersed in the chitin scaffolds before and after silicification. The size of silica particles used was observed to have a significant influence on the morphology of the obtained composites. The use of colloidal silica of micrometric size led to the homogeneous distribution of spherical $SiO_2$ particles on the surface of sponge chitin fibers. When colloidal silica with nanometric size particles was used, the chitinous scaffolds were completely coated with the inorganic material, this being also characterized by spherical morphology of nanoparticles.

For tissue engineering scaffolds, laminated porous structure is essential for allowing the diffusion of oxygen, nutrients and metabolic products to and from encapsulated cells (Huang et al. 2015). The ideal pore size range is 150–200 $\mu$m and hence SEM analysis can be used as an important tool for selecting bio scaffolds for tissue engineering. Jayakumar et al. (2011a, b) fabricated chitin–chitosan/nano $TiO_2$-composite scaffolds and chitin–chitosan/nano $ZrO_2$-composite scaffolds for tissue engineering applications. SEM images of the composite scaffolds showed that scaffolds were macro porous in nature. No visible changes were seen in chitin–chitosan, chitin–chitosan/nano$TiO_2$ composite scaffolds. Pore size of the composite scaffold varied from 150 to 200 $\mu$m as measured by SEM in comparison to the pore size of 200–300 $\mu$m of control chitin–chitosan scaffold and hence was found to be ideal for tissue engineering applications. SEM images was used to study the attachment, morphology and spreading of cells on the scaffolds. SEM images of cells incubated for 48 h on the scaffolds showed that cells attached and spread within the pore walls offered by the scaffolds. The result indicated that the nanocomposite scaffolds might be suitable for bone tissue engineering. The higher attachment on nanocomposite scaffolds may be due to increase in surface area. It is known that surface topology could play a role in cell attachment on implants. An increase in surface area allows maximum area for cell attachment and nanosurfaces have larger surface area to volume ratio. Surface morphology of β-chitin/nano silver was studied by Kumar et al. (2010). The composite scaffolds showed a highly porous and smooth surface. The highly porous structure of scaffold as compared to control chitin resulted from the presence of water in the silver colloid which when incorporated into the hydrogel, gets evaporated and the vacant space is left as pores in the hydrogel scaffold, ultimately resulting in highly porous composite scaffolds (Fig. 3.7). Morphology of chitin-hydroxyapatite composites showed that the composite structures have a heterogeneous porous structure. This information has a great influence on the final microstructure and degree of interconnectivity of the scaffolds (Silva et al. 2013).

To understand the role that chitin nanocrystals, nanofibers and nanowhiskers have on the structure and functional properties of thermoplastic films, these nano objects were incorporated at different fractions into polymer matrices and morphological analysis using different techniques like TEM, AFM and SEM were done. It was found that chitin nano particles in polymer matrices exhibited colloidal behavior, due to the protonation of the amino groups ($NH_3^+$) which induced positive charges on the surface of the crystallites and promoted the stability of the suspension. As the CNW content increased, a significant tendency of agglomeration was also observed.

Shankar et al. (2015) prepared chitin nanofibrils (CNF) reinforced carrageenan nanocomposite films by solution-casting technique. The nanocomposite films were

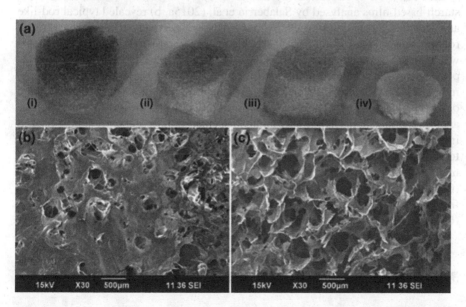

**Fig. 3.7** **a** Images of composite scaffolds. (i) β-Chitin + 0.006% nanosilver; (ii) β-chitin + 0.003% nanosilver; (iii) β-chitin + 0.001% nanosilver; (iv) β-chitin control. **b** SEM image of β-chitin control scaffold. **c** SEM image of β-chitin with nanosilver (0.006%). Reprinted with permission from Kumar et al. (2010). Copyright Elsevier

free-standing, flexible and homogeneous. The neat carrageenan film displayed smooth surface which indicated the formation of homogeneous film. For 3 and 5% of CNF uniform dispersion of CNF was observed in the polymer matrix. However, higher concentration of CNF (10 wt%) in carrageenan matrix resulted in non-uniform distribution and agglomeration of CNF, which led to the rough surface of films. The morphology of gelatin-based nanocomposite films containing 5 and 10% N-chitin was studied by Sahraee et al. (2017) and it was observed that gelatin formed a uniform and smooth surface film and in the case of applying N-chitin in its formulation, proper interactions occurred between the film matrix and nanoparticles due to their good compatibility. Here also it was seen through SEM analysis that when chitin nanoparticles were incorporated to gelatin films at low concentrations, they dispersed individually with less agglomeration but as the concentration increased chitin nanoparticle in the film matrix tended to aggregate.

Contrary to the above results, in the case of alginate hydrogels reinforced with chitin whisker content as high as 50 wt%, no obvious aggregation was observed, indicating good compatibility between the chitin whiskers and the alginate matrix (Huang et al. 2015). This was attributed to the pH-induced charge shifting behavior of chitin whiskers. When the pH was in the range of 11–12, the chitin whiskers could be well dispersed in the alginate matrix to form homogeneous architecture. Herein, the chitin whiskers retained nano sized distribution and formed entangled structure in the alginate matrix.

AFM analysis of chitin nanocrystals (CHNC) and nanofibers (CHNF) reinforced starch-based films analysed by Salaberria et al. (2015a, b) revealed typical rod-like and web-like morphology of CHNC and CHNF respectively. The average surface roughness expressed as the root mean squared roughness over an area of 25 $\mu m^2$ was found to be 63.8 nm for S/CHNC10 and 137 nm for S/CHNF10. The nanocomposite films were scanned at different locations and the profiles were very similar, indicating that the obtained materials presented a uniform surface of chitin nano objects overlapping each other. It was observed that due to the rod-like morphology and lower dimension, CHNC were highly dispersed and consequently solidly packed into thermoplastic starch matrix. Chitin nanowhiskers on the other hand was found to exhibit a needle-like morphology and as the CNW content increased in the starch matrix, a significant tendency of agglomeration was observed in the TEM images (Qin et al. 2016). Figure 3.8 shows the TEM images of the surfaces of maize starch films reinforced with 0.5% (a), 1% (b), 2% (c), and 5% (d) of chitin nanowhiskers (CNWs).

### 3.3.3  Morphological Analysis of Chitosan Based Composites

The morphological characteristics of PLA/CS/epoxidised natural rubber (ENR) composites were investigated by scanning electron microscopy (SEM) and optical microscopy By Zakaria et al. (2013). SEM analysis of composites fractured surfaces revealed smooth and homogeneous texture upon incorporation of CS in the PLA

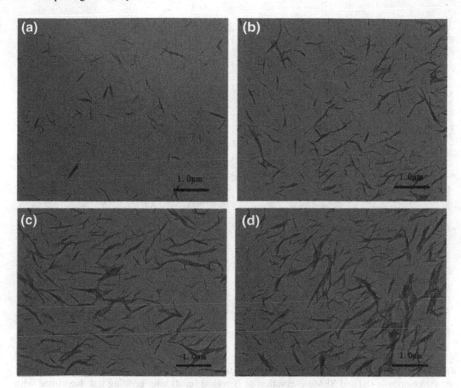

**Fig. 3.8** TEM images of the surfaces of maize starch films reinforced with 0.5% (**a**), 1% (**b**), 2% (**c**), and 5% (**d**) of chitin nanowhiskers (CNWs). Reprinted with permission from Qin et al. (2016). Copyright Elsevier

matrix due to the strong interaction between CS and PLA. However at higher loading (>10 wt%) the phase segregation and poor adhesion between the polymers were observed because of the van der Waals forces leading to a fiber bundle formation.

Hu et al. prepared a transparent yellow chitosan (CS)/(HA) nanocomposite for internal fixation of bone fracture (Hu et al. 2004). Homogeneous dispersion of HA particles into the CS matrix was confirmed by TEM analysis (Fig. 3.9). The SEM images of composites after the bending tests revealed the layer-by-layer structure of composite prepared by in situ hybridization.

The morphological analysis of $Fe_3O_4$/MWNT/Chitosan nanocomposite was done by FESEM (Marroquin et al. 2013). The fractured surface of these nanocomposites exhibited improved dispersion of MWNT characterized by the presence of individual isolated nanotubes within the chitosan matrix. A uniform distribution of MWNT was observed, with the ends of broken nanotubes on the fractured surface. Strong interfacial adhesion between MWNT and chitosan matrix was revealed by the examination, and most of the MWNT were broken rather than pulled out from the matrix (Fig. 3.10).

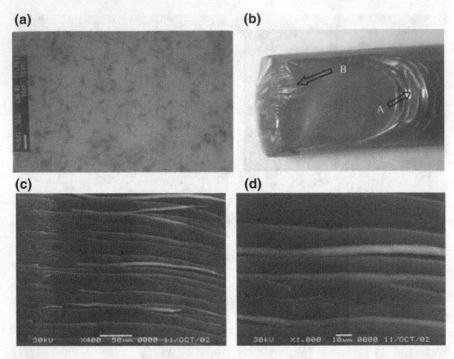

**Fig. 3.9** TEM micrographs of CS/HA (100/5, wt/wt) (**a** ×10K), photographs of CS/HA (100/5, wt/wt) composite (**b**) and SEM observation at B region (**c, d**) after bending test. Reprinted with permission from Qin et al. (2016). Copyright Elsevier

In another study, Archana et al. observed that owing to the viscous nature of chitosan–PVP complex matrix, the $TiO_2$ particles agglomerated in the matrix (Archana et al. 2013). The viscous nature of the matrix did not allow the $TiO_2$ nanoparticles to distribute uniformly in the whole binary polymer mixture under ambient stirring conditions. TEM studies (Fig. 3.11) showed spherical particles with good polycrystalline nature as shown by electron diffraction pattern (in inset). Figure 3.11d demonstrates the histogram of $TiO_2$ nanoparticles embedded in binary polymer matrix that have shown average particle size of nanoparticles of about 25–35 nm with a good distribution for short ranges.

The surface morphology of CS/MMT polymer composites by SEM analysis was studied by Thakur et al. (2016). They found that the smoothness of the CS surface was decrease with the addition of MMT and the interaction between CS and clay results in the presence of CS-rendered intercalated lamellar structures. Tan et al. pointed out the presence of stacked flakes in CS/MMT biocomposites as an evidence of the interaction between CS and MMT (Tan et al. 2008). The presence of lamellar structures in montmorillonite chitosan bionanocomposites were reported by Celis et al. (2012).

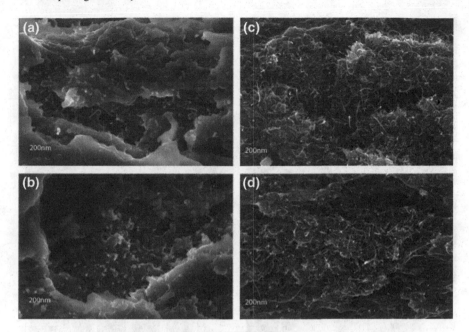

**Fig. 3.10** FESEM images of fracture surfaces of Chitosan nanocomposites. Reprinted with permission from Marroquin et al. (2013). Copyright Elsevier

Habiba et al. (2017) studied the morphology of electrospun chitosan/PVA/Zeolite composites and found that zeolite particles were embedded into the nanofibers. Particles were uniformly distributed over the nanofiber surface. The uniform particle distribution as well as formation of rough and porous surface was attributed to the strong interaction of zeolite with the functional group of chitosan and PVA. At a higher filler concentration, they observed an increase in fiber diameter due to higher viscoelastic force which caused the higher resistance toward electrostatic force that stretched the jet. Surface morphological analysis of chitosan/polyvinyl alcohol/zeolite composite was conducted by Habiba et al. (2017) to determine the dispersion of zeolite over the chitosan/PVA matrix. A rough and porous surface was reported which led to high hydrophilicity and hence influencing the adsorption efficiency. Filler particles were agglomerated in the polymer matrix probably because of the van der Waals forces that induce a bundle formation of these materials. Therefore, a poor interfacial adhesion occurred between the polymer matrix and fiber.

## 3.4 Barrier Properties

In food packaging, a coating or a film is often required to decrease or avoid the moisture transfer between the food and the neighbouring atmosphere. Owing to the

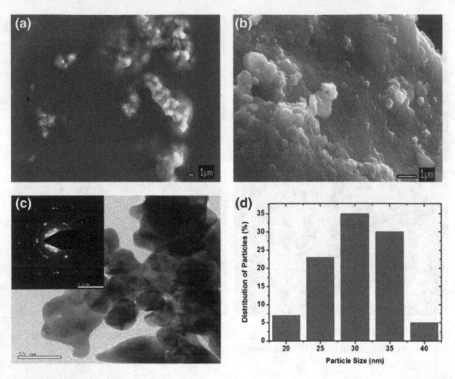

**Fig. 3.11** Microscopic images of the chitosan–PVP–TiO$_2$ dressing material: scanning electron micrograph depicting a homogenous dispersion of titanium dioxide nanoparticles (scale 11 mm) (**a**, **b**), TEM of composite dressing material (scale 50 nm) (**c**) and histogram revealing the nanoparticle size distribution (**d**). Reprinted with permission from Archana et al. (2013). Copyright Elsevier

direct influence on the deteriorative reactions in packed food products, water vapor permeability (WVP) is the most important and extensive property of bio-based polymer films. The high WVP of most bio-based polymers makes them inappropriate for several applications. The barrier properties can be improved by the use of additives.

The addition of chitin whisker was found to significantly reduce the water absorption property of the matrix film. The decrease was attributed to the effect of impermeable crystalline chitin molecules which provided a tortuous path of water vapor diffusion through the polymer matrix. In the case of carrageenan/CNF nanocomposite films moisture content of carrageenan films decreased significantly after blending with CNF, and the moisture content decreased linearly up to 5 wt% of CNF, then increased with higher content of CNF (10 wt%). This indicated that the impermeable crystalline chitin molecules dispersed well in the carrageenan polymer matrix. However, when more than 5 wt% of CNF was included, they exhibited aggregation resulting in decreased effective filler content and facilitated water vapour permeation.

The WVP of maize starch nanocomposite films decreased with chitin nanowhisker (CNW) content increasing from 0 to 2%. However, higher concentration of CNWs

(5%) aggregated easily, which actually decreased the effective content of the CNWs and facilitated water vapour permeation. At low CNW concentrations the CNWs dispersed well in the matrix, and thus, there were fewer paths for water molecules to pass through (Qin et al. 2016). Regarding starch/chitin nanocrystals (CHNC) nanocomposite films, when 5 and 10 wt% of CHNC were added, vapour transmission rate (WVTR) values decreased. On the other hand, the S/CHNF nanocomposite films only showed better water resistance for the materials prepared with 5 wt% of CHNF. In addition to agglomeration property of chitin, another reason for the poor water vapour barrier property is the presence of excessive residual $NH_2$ groups, which present more affinity for water than OH groups, at the surface of the chitin nano-objects (Salaberria et al. 2015a, b).

Chitosan nanocomposites with MgO nanoparticles showed superior UV-shielding effect (De Silva et al. 2017). This is due to the UV-shielding ability of MgO nanoparticles. The moisture barrier properties also enhanced with nano MgO. These improvements could be related to the presence of dispersed MgO nanoparticles within the matrix which would act as a barrier to hinder the diffusion of water vapor through the nanocomposites and MgO could absorb hydroxyl groups to form $Mg(OH)_2$ during the water vapor permeation.

# References

Archana D, Singh BK, Dutta J, Dutta PK (2013) In vivo evaluation of chitosan–PVP-titanium dioxide nanocomposite as wound dressing material. Carbohydr Polym 95:530–539. https://doi.org/10.1016/j.carbpol.2013.03.034

Cao X, Chen Y, Chang PR, Muir AD, Falk G (2008) Starch-based nanocomposites reinforced with flax cellulose nanocrystals. Express Polym Lett 2:502–510. https://doi.org/10.3144/expresspolymlett.2008.60

Celis R, Adelino MA, Hermosín MC, Cornejo J (2012) Montmorillonite–chitosan bionanocomposites as adsorbents of the herbicide clopyralid in aqueous solution and soil/water suspensions. J Hazard Mater 209:67–76. https://doi.org/10.1016/j.jhazmat.2011.12.074

Chang PR, Jian R, Yu J, Ma X (2010) Starch-based composites reinforced with novel chitin nanoparticles. Carbohydr Polym 80:421–426. https://doi.org/10.1016/j.carbpol.2009.11.041

Chen C, Li D, Deng Q, Zheng B (2012) Optically transparent biocomposites: polymethylmethacrylate reinforced with high-performance chitin nanofibers. BioResources 7:5960–5971. https://doi.org/10.15376/biores.7.4.5960-5971

De Silva RT, Mantilaka MM, Ratnayake SP, Amaratunga GA, de Silva KN (2017) Nano-MgO reinforced chitosan nanocomposites for high performance packaging applications with improved mechanical, thermal and barrier properties. Carbohyd Polym 157:739–747. https://doi.org/10.1016/j.carbpol.2016.10.038

Gironès J, López JP, Mutjé P, Carvalho AJFD, Curvelo AADS, Vilaseca F (2012) Natural fiber-reinforced thermoplastic starch composites obtained by melt processing. Compos Sci Technol 72:858–863. https://doi.org/10.1016/j.compscitech.2012.02.019

Habiba U, Afifi AM, Salleh A, Ang BC (2017) Chitosan/(polyvinyl alcohol)/zeolite electrospun composite nanofibrous membrane for adsorption of $Cr^{6+}$, $Fe^{3+}$ and $Ni^{2+}$. J Hazard Mater 322:182–194. https://doi.org/10.1016/j.jhazmat.2016.06.028

Harfiz M, Salleh E, Nur S et al (2014) Starch based active packaging film reinforced with empty fruit bunch (EFB) cellulose nanofiber. Procedia Chem 9:23–33. https://doi.org/10.1016/j.proche. 2014.05.004

Herrera N, Salaberria AM, Mathew AP, Oksman K (2016) Plasticized polylactic acid nanocomposite films with cellulose and chitin nanocrystals prepared using extrusion and compression molding with two cooling rates: effects on mechanical, thermal and optical properties. Compos A Appl Sci Manuf 83:89–97. https://doi.org/10.1016/j.compositesa.2015.05.024

Hu Q, Li B, Wang M, Shen J (2004) Preparation and characterization of biodegradable chitosan/hydroxyapatite nanocomposite rods via in situ hybridization: a potential material as internal fixation of bone fracture. Biomaterials 25:779–785. https://doi.org/10.1016/S0142-9612(03)00582-9

Huang Y, Yao M, Zheng X, Liang X, Su X, Zhang Y, Lu A, Zhang L (2015) Effects of chitin whiskers on physical properties and osteoblast culture of alginate based nanocomposite hydrogels. Biomacromol 16:3499–3507. https://doi.org/10.1021/acs.biomac.5b00928

Jayakumar R, Ramachandran R, Divyarani VV, Chennazhi KP, Tamura H, Nair SV (2011a) Fabrication of chitin-chitosan/nano TiO$_2$-composite scaffolds for tissue engineering applications. Int J Biol Macromol 48:336–344. https://doi.org/10.1016/j.ijbiomac.2010.12.010

Jayakumar R, Ramachandran R, Sudheesh Kumar PT, Divyarani VV, Chennazhi KP, Tamura H, Nair SV (2011b) Fabrication of chitin-chitosan/nano ZrO$_2$ composite scaffolds for tissue engineering applications. Int J Biol Macromol 49:274–280. https://doi.org/10.1016/j.ijbiomac.2011.04.020

Jin J, Hassanzadeh P, Perotto G, Sun W, Brenckle MA, Kaplan D, Omenetto FG, Rolandi M (2013) A biomimetic composite from solution self-assembly of chitin nanofibers in a silk fibroin matrix. Adv Mater 25:4482–4487. https://doi.org/10.1002/adma.201301429

Junkasem J, Rujiravanit R, Supaphol P (2006) Fabrication of α-chitin whisker-reinforced poly(vinyl alcohol) nanocomposite nanofibres by electrospinning. Nanotechnology 17:4519–4528. https://doi.org/10.1088/0957-4484/17/17/039

Kadokawa J, Takegawa A, Mine S, Prasad K (2011) Preparation of chitin nanowhiskers using an ionic liquid and their composite materials with poly(vinyl alcohol). Carbohydr Polym 84:1408–1412. https://doi.org/10.1016/j.carbpol.2011.01.049

Kumar PTS, Abhilash S, Sreeja V, Tamura H, Manzoor K, Nair SV, Jayakumar R (2010) Development of novel chitin/nanosilver composite scaffolds for wound dressing applications. J Mater Sci Mater Med 21:807–813. https://doi.org/10.1007/s10856-009-3877-z

Kumar PT, Lakshmanan VK, Anilkumar TV, Ramya C, Reshmi P, Unnikrishnan AG, Nair SV, Rs Jayakumar (2012) Flexible and microporous chitosan hydrogel/nano ZnO composite bandages for wound dressing: in vitro and in vivo evaluation. ACS Appl Mater Interfaces 4:2618–2629. https://doi.org/10.1021/am300292v

Li R, Liu C, Ma J (2011) Studies on the properties of graphene oxide-reinforced starch biocomposites. Carbohydr Polym 84:631–637. https://doi.org/10.1016/j.carbpol.2010.12.041

Lopez O, Garcia M, Villar M, Gentili A, Rodriguez MS, Albertengo L (2014) Thermo-compression of biodegradable thermoplastic corn starch films containing chitin and chitosan. LWT - Food Sci Technol 57:106–115. https://doi.org/10.1016/j.lwt.2014.01.024

Lu Y, Weng L, Zhang L (2004) Morphology and properties of soy protein isolate thermoplastics reinforced with chitin whiskers. Biomacromol 5:1046–1051. https://doi.org/10.1021/bm034516x

Marroquin JB, Rhee KY, Park SJ (2013) Chitosan nanocomposite films: enhanced electrical conductivity, thermal stability, and mechanical properties. Carbohydr Polym 92:1783–1791. https://doi.org/10.1016/j.carbpol.2012.11.042

Nascimento P, Marim R, Carvalho G, Mali S (2016) Nanocellulose produced from rice hulls and its effect on the properties of biodegradable starch films. Mater Res 19(1):167–174. https://doi.org/10.1590/1980-5373-MR-2015-0423

Neto CD, Giacometti JA, Job AE, Ferreira FC, Fonseca JL, Pereira MR (2005) Thermal analysis of chitosan based networks. Carbohydr Polym 62:97–103. https://doi.org/10.1016/j.carbpol.2005. 02.022

Orue A, Corcuera MA, Pena C, Eceiza A, Arbelaiz A (2014) Bionanocomposites based on thermoplastic starch and cellulose nanofibers. J Thermoplast Compos Mater 29:817–832. https://doi.org/10.1177/0892705714536424

Pandey JK, Chu WS, Kim CS, Lee CS, Ahn SH (2009) Bio-nano reinforcement of environmentally degradable polymer matrix by cellulose whiskers from grass. Compos B Eng 40:676–680. https://doi.org/10.1016/j.compositesb.2009.04.013

Qin Y, Zhang S, Yu J, Yang J, Xiong L, Sun Q (2016) Effects of chitin nano-whiskers on the antibacterial and physicochemical properties of maize starch films. Carbohydr Polym 147:372–378. https://doi.org/10.1016/j.carbpol.2016.03.095

Ramírez MGL, Satyanarayana KG, Iwakiri S, de Muniz GB, Tanobe V, Flores-Sahagun TS (2011) Study of the properties of biocomposites. Part I. Cassava starch-green coir fibers from Brazil. Carbohydr Polym 86:1712–1722. https://doi.org/10.1016/j.carbpol.2011.07.002

Rubentheren V, Ward TA, Chee CY, Tang CK (2015) Processing and analysis of chitosan nanocomposites reinforced with chitin whiskers and tannic acid as a crosslinker. Carbohydr Polym 115:379–387. https://doi.org/10.1016/j.carbpol.2014.09.007

Sahraee S, Milani JM, Ghanbarzadeh B, Hamishehkar H (2017) Physicochemical and antifungal properties of bio-nanocomposite film based on gelatin-chitin nanoparticles. Int J Biol Macromol 97:373–381. https://doi.org/10.1016/j.ijbiomac.2016.12.066

Salaberria AM, Labidi J, Fernandes SCM (2014) Chitin nanocrystals and nanofibers as nano-sized fillers into thermoplastic starch-based biocomposites processed by melt-mixing. Chem Eng J 256:356–364. https://doi.org/10.1016/j.cej.2014.07.009

Salaberria AM, Diaz RH, Labidi J, Fernandes SCM (2015a) Role of chitin nanocrystals and nanofibers on physical, mechanical and functional properties in thermoplastic starch films. Food Hydrocoll 46:93–102. https://doi.org/10.1016/j.foodhyd.2014.12.016

Salaberria AM, Labidi J, Fernandes SCM (2015b) Different routes to turn chitin into stunning nano-objects. Eur Polym J 68:503–515. https://doi.org/10.1016/j.eurpolymj.2015.03.005

Shankar S, Reddy JP, Rhim JW, Kim HY (2015) Preparation, characterization, and antimicrobial activity of chitin nanofibrils reinforced carrageenan nanocomposite films. Carbohydr Polym 117:468–475. https://doi.org/10.1016/j.carbpol.2014.10.010

Silva SS, Duarte ARC, Oliveira JM (2013) Alternative methodology for chitin-hydroxyapatite composites using ionic liquids and supercritical fluid technology. J Bioact Compat Polym 28:481–491. https://doi.org/10.1177/0883911513501595

Souza AC, Goto GEO, Mainardi JA, Coelho ACV, Tadini CC (2013) Cassava starch composite films incorporated with cinnamon essential oil: antimicrobial activity, microstructure, mechanical and barrier properties. LWT - Food Sci Technol 54:346–352. https://doi.org/10.1016/j.lwt.2013.06.017

Tan W, Zhang Y, Szeto YS, Liao L (2008) A novel method to prepare chitosan/montmorillonite nanocomposites in the presence of hydroxy-aluminum oligomeric cations. Compos Sci Technol 68:2917–2921. https://doi.org/10.1016/j.compscitech.2007.10.007

Thakur G, Singh A, Singh I (2016) Chitosan-montmorillonite polymer composites: formulation and evaluation of sustained release tablets of aceclofenac. Sci Pharm 84:603–617. https://doi.org/10.3390/scipharm84040603

Uddin AJ, Fujie M, Sembo S, Gotoh Y (2012) Outstanding reinforcing effect of highly oriented chitin whiskers in PVA nanocomposites. Carbohydr Polym 87:799–805. https://doi.org/10.1016/j.carbpol.2011.08.071

Wang J, Wang Z, Li J, Wang B, Liu J, Chen P, Miao M, Gu Q (2012) Chitin nanocrystals grafted with poly(3-hydroxybutyrate-co-3-hydroxyvalerate) and their effects on thermal behavior of PHBV. Carbohydr Polym 87:784–789. https://doi.org/10.1016/j.carbpol.2011.08.066

Watthanaphanit A, Supaphol P, Tamura H, Tokura S, Rujiravanit R (2008) Fabrication, structure, and properties of chitin whisker-reinforced alginate nanocomposite fibers. J Appl Polym Sci 110:890–899. https://doi.org/10.1002/app

Wysokowski M, Behm T, Born R, Bazhenov VV, Meißner H, Richter G, Szwarc-Rzepka K, Makarova A, Vyalikh D, Schupp P, Jesionowski T (2013a) Preparation of chitin–silica composites by in vitro silicification of two-dimensional Ianthella basta demosponge chitinous scaffolds under modified Stöber conditions. Mater Sci Eng C 33:3935–3941. https://doi.org/10.1016/j. msec.2013.05.030

Wysokowski M, Motylenko M, Stöcker H, Bazhenov VV, Langer E, Dobrowolska A, Czaczyk K, Galli R, Stelling AL, Behm T, Klapiszewski L (2013b) An extreme biomimetic approach: hydrothermal synthesis of β-chitin/ZnO nanostructured composites. J Mater Chem B 1:6469–6476. https://doi.org/10.1039/c3tb21186j

Wysokowski M, Motylenko M, Beyer J, Makarova A, Stöcker H, Walter J, Galli R, Kaiser S, Vyalikh D, Bazhenov VV, Petrenko I (2015a) Extreme biomimetic approach for developing novel chitin-GeO$_2$ nanocomposites with photoluminescent properties. Nano Res 8:2288–2301. https://doi. org/10.1007/s12274-015-0739-5

Wysokowski M, Petrenko I, Motylenko M, Langer E, Bazhenov VV, Galli R, Stelling AL, Kljajić Z, Szatkowski T, Kutsova VZ, Stawski D (2015b) Renewable chitin from marine sponge as a thermostable biological template for hydrothermal synthesis of hematite nanospheres using principles of extreme biomimetics. Bioinspired Mater 1:12–22. https://doi.org/10.1515/bima-2015-0001

Zakaria Z, Islam MS, Hassan A, Mohamad Haafiz MK, Arjmandi R, Inuwa IM, Hasan M (2013) Mechanical properties and morphological characterization of PLA/chitosan/epoxidized natural rubber composites. Adv Mater Sci Eng 2013(629092). https://doi.org/10.1155/2013/629092

# Chapter 4
# Applications of Polysaccharide Based Composites

## 4.1 Packaging Applications

The perfect packaging material with excellent mechanical and barrier properties and biodegradability are obtained from renewable biological resources, usually called biopolymers. Biopolymer packaging materials also serve as gas and solute barriers and challenge other types of packaging by improving the quality and extending the shelf-life of foods. The properties of these biodegradable materials can be enhanced by incorporating a wide variety of additives, such as antioxidants, antifungal agents, antimicrobials, colors, and other nutrients (Rhim et al. 2013).

In order to develop active packaging, cinnamon was incorporated into cassava starch. The effect of cinnamon essential oil on antimicrobial activity, mechanical and barrier properties of films was evaluated and the results were compared with those of control films (without antimicrobial agent). They showed that the essential oil content significantly influenced the properties of the composite films. A great number of studies on the antimicrobial characteristics of films made from starch have been carried out earlier. Nevertheless, no information has been presented about the effect of cinnamon essential oil on *P. commune* and *E. amstelodami*, which plays an important role in the spoilage of bread products. The results established that cassava starch films can be considered as a potential active alternative packaging material, while further research is necessary to improve their mechanical and barrier properties since adequate mechanical properties are generally required for a packaging film to withstand external stress and maintain its integrity as well as barrier properties during applications as food packaging (Souza et al. 2013). Figure 4.1 shows the image of starch films incorporated with cinnamon essential oil.

Harfiz et al. prepared cellulose nanofiber reinforced starch based composite films for active packaging. Cellulose nanofiber was extracted from oil palm empty fruit bunch (OPEFB). Figure 4.2 shows the photographs of active packaging films (Harfiz et al. 2014). The nanofibers were found to be appropriate filler for thin film packaging because they had not strongly affected the appearance of the film compared to the

**Fig. 4.1** Cassava starch film incorporated with 0.4 g of cinnamon essential oil/100 g of filmogenic solution %). Reprinted with permission from Souza et al. (2013). Copyright Elsevier

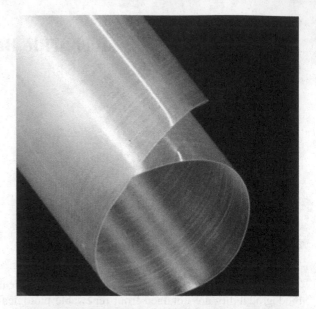

**Fig. 4.2** Starch based active packaging film, 0, 2, 4, 6, 8 and 10% of cellulose nanofiber incorporation. Reprinted with permission from Harfiz et al. (2014). Copyright Elsevier

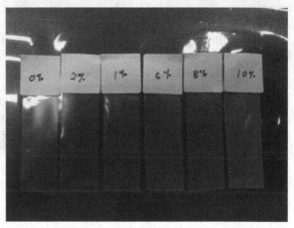

micro size fibers which create a lumpy structure deposited on the surface of the film (Harfiz et al. 2014).

Chitin can be applied in edible film industries, for instance, in controlling of moisture transfer between food and surroundings, heat transfer, rate of respiration, release of antioxidants and also enzymatic browning in fruits. Chitin/chitosan whisker rectorite ternary films prepared (Li et al. 2011) was found to have higher thermal stability and smoother surface in comparison with chitin whisker or rectorite-reinforced binary films. Due to good properties such as water resistance and anti-bacterial activity, this ternary film also had high potential in food-packaging applications. Chitin nano crystal (CNC) and whey protein isolate (WPI) aqueous dispersions prepared (at pH 3) (Tzoumaki et al. 2011) showed that by increasing CNC or WPI concentration, the CNC-WPI mixed dispersion showed a gel-like behaviour and this material was found to have a potential in new food product formulations and as composite matrix for bioactive delivery.

Qin et al. (2016) and Shankar et al. (2015) prepared maize starch/CNW nanocomposite films carrageenan/CNF nanocomposite film respectively with improved properties to prolong the shelf life of packaged foods. The composite films showed improved thermal properties and mechanical strength, and conferred high antibacterial activity. The nanocomposite films exhibited stronger antimicrobial activity against Gram-positive *L. monocytogenes* than against Gram-negative *E. coli*. CHNC and CHNF materials were added as reinforcing agents in thermoplastic starch films for food packaging by Salaberria et al. (2015a, b). The effects induced by CHNF on the final properties of thermoplastic starch matrix were considerably higher than those CNC due to the web-like morphology of CHNF.

Chitosan nanocomposite thin films were fabricated by incorporating MgO nanoparticles to significantly improve its physical properties for potential packaging applications (De Silva et al. 2017). The chitosan/MgO nanocomposites showed remarkable thermal stability, flame retardant properties, UV shielding and moisture barrier properties, hence they are appropriate for food packaging applications.

One of the greatest challenges in the field of food packaging is microbial contamination. Hence, Al-Naamani et al. (2016) prepared a polyethylene coated with chitosan-ZnO nanocomposite which provides enhanced antimicrobial properties of the films in order to prolong the shelf life of food products. Chitosan-ZnO nanocomposite coatings improved antibacterial properties of polyethylene by deactivating about 99.9% of viable pathogenic bacteria.

Fish gelatin/chitosan nanoparticle based bioactive films containing oregano essential oil were developed by Hosseini et al. (2016). Addition of essential oil resulted in more flexible films, with a decrease in water vapor permeability and exhibited distinctive antimicrobial activity.

## 4.2 Water Treatment

Addition of chemicals into water system is required for almost all conventional water treatment processes. This includes addition of acids and bases (pH adjustment), flocculating agents, coagulants and so on which usually generate new problems. Besides, conventional processes are not effective enough. For example, heavy metal ions cannot be removed by precipitation if its ionic product is lower than the solubility product (Liu et al. 2014). Chitin has been reported to be useful in heavy metal chelating of industrial wastewater. Chitin was found to bind common water pollutants like mercury, copper, iron, nickel, chromium, lead, zinc, cadmium, silver, and cobalt. The results showed that the strongest binding took place with mercury and the weakest with cobalt.

Chitin hydrogel/$SiO_2$ hybrid and chitosan hydrogel/$SiO_2$ hybrid have been prepared using sol–gel technique to be used as biosorbents (Copello et al. 2011) and these were investigated for adsorption of four dyes (Remazol Black B, Erythrosine B, Neutral Red and Gentian Violet). The results showed that aside from highly charged dyes, the chitin containing matrix had similar or higher adsorption capacity than their chitosan counterparts.

A novel chitin coated with polyaniline (PCC) has been successfully synthesized, characterized and applied for the removal of Cr(VI) ion from aqueous solution by batch mode (Karthik and Meenakshi 2015). The removal of Cr(VI) ions by the PCC was found to be highly pH dependent and was mainly driven by electrostatic adsorption coupled reduction.

Chitin/cellulose composite membranes with micro porous structures exhibited good efficiency for removal of heavy metal ions such as mercury, copper and lead (Tang et al. 2011). The results revealed that uptake capacity of the heavy metal ions increased with increasing chitin content. The major mechanism of the adsorption of metal ions by chitin/cellulose composite membranes included metal chelation and ion exchange. Moreover, the chitin/cellulose membranes could be easily regenerated. This work provided a "green" pathway for the removal of hazardous materials in waste water (Tang et al. 2011).

Chitosan can be used as an adsorbent to remove heavy metals and dyes due to the presence of amino and hydroxyl groups, which can serve as active sites. Amino groups of chitosan are capable to adsorb anionic dyes by electrostatic attraction in the acidic media, after the cationization of amino groups. The activity of chitosan can be improved by crosslinking it with glyoxal, formaldehyde, glutaraldehyde, epichlorohydrin etc., or by modifications such as carboxylation, amination, incorporation of magnetic particles, hydroxyapatites, multi–walled carbon nanotubes, montmorillonite, etc. (Ngah et al. 2011).

The adsorption capacity of porous polymer matrix can be enhanced by incorporating magnetic nanoparticles into the matrix by way of ionic interactions, ion exchange, metal chelation and formation of ion–pairs (Bhatnagar and Sillanpää 2009). Nano sized iron oxides like magnetite ($Fe_3O_4$) and maghemite ($\gamma$-$Fe_2O_3$) can be modified to achieve better magnetic properties, lower toxicity and lower price. Tran et al. (2010) reported that hydrogel (2-acrylamido-2-methyl-1-propansulfonic acid, AMPS) cross-linked with chitosan matrix and magnetite particles were more attractive for the removal of pollutants like Pb(II) and Ni(II). Huang et al. (2009) investigated the adsorption of Cr(VI) on chitosan–magnetite composites using epichlorohydrin as the crosslinking agent.

The adsorption mechanism of graphene oxide–based magnetic cyclodextrin—chitosan nanocomposite is an example of either electrostatic interaction between negatively charged $HCrO_4{}^-$ with protonated ammonium (R–$NH_3{}^+$) ion or CCGO $OH_2{}^+$, or redox reaction which includes the reduction of $Cr^{6+}$ to $Cr^{3+}$ ions (Liu et al. 2014). But in another study Sivakami et al. pointed out that the reasonable mechanism for the interactions of nanochitosan–chromium(VI) is electrostatic attraction between oppositely charged ions (Sivakami et al. 2013).

Introduction of nano-hydroxyapatite (n-HA) into the biopolymeric matrix improved its mechanical strength and provided stability to the polymer. Hence such composites can be widely used for heavy metal adsorption. Gandhi et al. proposed that ion-exchange mechanism is responsible for the removal of Cu(II) by Chitin/chitosan n-HA composites (Gandhi et al. 2011).

A chitosan/PVA/zeolite electrospun composite was found to be appropriate for the removal of Cr(VI), Fe(III) and Ni(II) (Habiba et al. 2017). The results suggested that the electrospun composites exhibited versatility in metal removal at medium concentration of heavy metals and adsorption capacity of nanofiber was unaltered after five recycling runs. Therefore, the chitosan/PVA/zeolite composite ensures the reusability of membrane for water treatment.

Dye removal mechanism of nanochitosan also involves a range of processes like chemical bonding, ion exchange, hydrogen bonds, hydrophobic attractions, van der Waals force, physical adsorption, aggregation mechanisms, dye interactions etc., which can take place simultaneously (Crini and Badot 2008). The electrostatic interaction between negatively charged dye ions and protonated amino groups of chitosan is responsible for the adsorption mechanism of acid dyes with nano chitosan (Zhou et al. 2011). Shen et al. (2011) proposed that adsorption, desorption and regeneration are the main steps involved in dye removal and explained that when dyes are to be adsorbed from alkaline effluent, the adsorption mechanism involved chelating interactions rather than electrostatic interactions.

The uptake of Congo red dye on n-HA/chitosan composite was investigated by Hou et al. (2012) and they noted a uniform dispersion of n-HA in the composite matrix. Electrostatic interaction, surface complexation, ion exchange and hydrogen bonding interaction between chitosan and Congo red are the possible pathways of adsorption.

Chitosan nanocomposites based on montmorillonite (CS/MMT) have been investigated for the removal of Congo red by Wang and Wang (2007). Results showed that CS/MMT nanocomposites exhibited improved adsorption capacity than the average values of that of neat CS and MMT at any pH. They proposed two possible pathways of adsorption mechanism : (a) electrostatic attraction between protonated chitosan and anionic dye and (b) the chemical reaction between Congo red and CS/MMT composite (Wang and Wang 2007). A reasonable mechanism of biosorption is shown in Fig. 4.3.

## 4.3  Biosensor

A biosensor is an analytical device that uses specific biochemical reactions mediated by a biological recognition. Chitosan can be employed as an apt matrix material for immobilizing bioactive molecules and constructing biosensors due to the presence of a large number of amino groups, good biocompatibility, simplicity of chemical modifications, superior permeability to water, biodegradability, antibacterial properties, and excellent film-forming ability originating from its protonation and solubility in

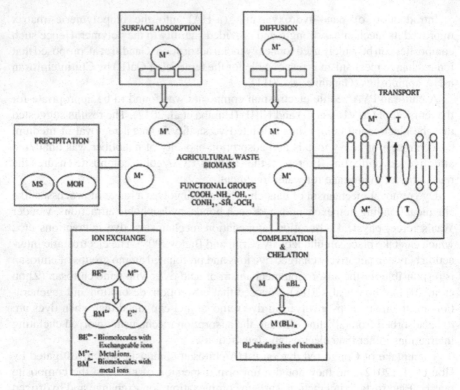

**Fig. 4.3** Reasonable mechanism of biosorption. Reprinted with permission from Olivera et al. (2016). Copyright Elsevier

slightly acidic solution (Liu et al. 2014). A schematic presentation of a chitosan-based biosensor is shown in Fig. 4.4.

A glucose biosensor based on immobilization of glucose oxidase in thin films of chitosan containing nanocomposites of graphene and gold nanoparticles (AuNPs) in a gold electrode was developed by Shan et al. (2010). The composites film showed prominent electrochemical response to glucose and the synergistic effect of graphene and AuNPs may promote the electro catalysis toward hydrogen peroxide. The high sensitivity and good stability at such a modified electrode led to glucose biosensor. A composite film of Au @CDs–CS/GCE electrode with high sensitivity, good specificity and stability were prepared by Huang et al. (2014) and can be used for the reliable determination of dopamine.

Nanocomposite thin films of ZnO nanoparticles and chitosan fabricated to immobilize granular porous morphology of Nano ZnO-CS to provide a better biocompatible environment for the enzyme. Here zinc oxide nanoparticles (NanoZnO) uniformly dispersed in chitosan was used to fabricate a hybrid nanocomposite film onto indium-tin-oxide (ITO) glass plate. Cholesterol oxidase (ChOx) was immobilized onto this NanoZnO–CS composite film using physisorption technique. Cholesterol oxidase/NanoZnO chitosan/ITO bioelectrode has been used for estimation of cholesterol in solution as well as in blood serum samples.

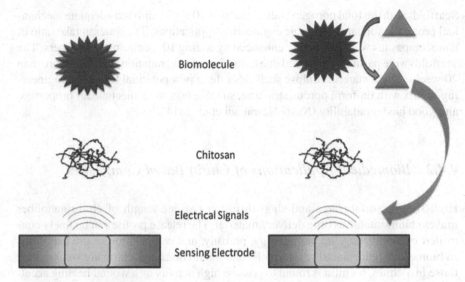

**Fig. 4.4** Schematic presentation of chitosan-based biosensor. Reprinted with permission from Shukla et al. (2013). Copyright Elsevier

## 4.4 Biomedical Application

### 4.4.1 Biomedical Applications of Starch Based Composites

Raj and Prabha (2016) studied the feasibility of Cassava starch acetate (CSA)–polyethylene glycol (PEG)–gelatin (G) nanocomposites as controlled drug delivery systems. It is one of the novel drug vehicles which can be used for the controlled release of an anticancer drug. A simple nano precipitation method was used to prepare the carriers CSA–PEG–G nanocomposites which were used for entrapping cisplatin (CDDP). The nanocomposites showed pH and time dependent drug release as confirmed by the in vitro drug dissolution profiles. Drug penetration and in vitro tests suggest that further studies are required to develop in vivo drug delivery systems. These results suggest that the CDDP coated CSA, CSA–PEG and CSA–PEG–G nanocomposites might be used as carriers for controlled drug delivery system.

Tissue engineering is an effective and confident method to repair or reestablish injured or destroyed organs. In this approach, a scaffold is fabricated in a three-dimensional medium suitable to seed and develop cells. Normally, cells are developed on a biodegradable porous scaffold in which the tissue is forms and grows, and the scaffold gradually degrades.

The porous thermoplastic starch-based composites prepared by combining film casting, salt leaching and freeze drying methods can be used as scaffolds. Composites had interconnected porous morphology; however an increase in pore interconnectivity was observed when the sodium chloride ratio was increased in the salt leaching.

Scaffolds with the total porogen (salt) content of 70 wt% exhibited adequate mechanical properties for cartilage tissue engineering applications. The water uptake ratio of nanocomposites was remarkably enhanced by adding 10% cellulose nanofibers. The scaffolds were partially destroyed due to low in vitro degradation rate after more than 20 weeks. These starch/cellulose scaffolds offer a new potential for tissue engineering matrix with uniform porous structure, suitable pore size, mechanical properties, and good biodegradability (Nasri-Nasrabadi et al. 2014).

### 4.4.2  Biomedical Applications of Chitin Based Composites

High specific surface area and short diffusion passage length of chitin nanofiber makes chitin suitable as drug delivery material. The release profile can be finely controlled by modulating the morphology, porosity, and composition of the nanofiber. In biomedical field, since the intermolecular interactions in $\beta$-chitin are weaker than those in $\alpha$-chitin, $\beta$-chitin is found to possess high activity as a wound healing accelerator (Usami et al 1998). Kumar et al. (2010) prepared novel $\beta$-chitin/nano silver composite scaffolds for wound healing applications with good mechanical strength, flexibility, tear resistance and excellent antimicrobial properties. Good retention of moisture environment, permeation of gas, absorption of wound exudates and easy removal without trauma to the wound are some of the advantages of using chitin scaffolds. Chitin–chitosan/nano $TiO_2$ composite scaffolds for bone tissue engineering were prepared by Jayakumar et al. (2011a, b). Results indicated no sign of toxicity and cells were found attached to the pore walls within the scaffolds. Novel $\beta$-chitin/ZnO film-like composites were prepared for the first time via hydrothermal synthesis by Wysokowski et al. (2013a, b) and they found that the prepared chitin/ZnO composite possesses antibacterial properties against Gram positive bacteria, which gives them good prospects in development of chitin based inorganic–organic wound dressing materials.. Nanocomposite scaffold of chitin–chitosan with nano $ZrO_2$ was fabricated by lyophilization technique using by Jayakumar et al. (2011a, b) and it was found that the incorporation of nano $ZrO_2$ onto the chitin–chitosan scaffold enhanced osteogenesis. Novel nanocomposite hydrogels composed of polyelectrolytes alginate and chitin whiskers with biocompatibility were successfully fabricated by Huang et al. (2015). The nanocomposite hydrogels exhibited similar crystallinity of 86.6% to the original chitin nanofiber (crystallinity: 92.5%). Furthermore, the introduction of chitin whiskers greatly improved the cytocompatibility of the composite hydrogel. With an increase of the chitin whisker content, the adhesion, spreading, and proliferation of osteoblast MC3T3-E1 on the composite hydrogels were significantly enhanced. Curcumin loaded chitin nanogels (CCNGs) were developed and evaluated (Mangalathillam et al. 2012), which can be used for treatment of melanoma through a transdermal route. These CCNGs possess desirable size and surface properties, excellent capacity for drug loading and release and good skin penetration and retention properties. Additionally, these chitin substrates are optically transparent and may find use in retinal regeneration (Hassanzadeh et al. 2013). Recently, tridimensional chitin/polyurethane network and chitin/polyurethane blends have been prepared. Both materials have high stability, low mass loss (in vivo) and also no release of toxic material and low adhesion to Vero cells. A number of studies have been reported where chitin films and membranes were used to treat patients with deep burns and with orthopedic injury.

### 4.4.3  Biomedical Application of Chitosan Based Composites

Chitosan is a valuable polymer for biomedical applications due to its biocompatibility, biodegradability and low toxicity. Chitosan is also used as sponges and bandages for the treatment of wounds, due to its biocompatible and biodegradable nature. But its applications are limited due to its insolubility in water and low reactivity (Zargar et al. 2015).

Biodegradability of chitin and chitosan is principally attributed to their vulnerability to enzymatic hydrolysis by lysozyme, a non-specific proteolytic enzyme that exists in all human body tissues. Lipase, an enzyme which exists in saliva and human gastric and pancreatic fluids, has the ability to degrade chitosan (Pantaleone et al. 1992). Chitosan stimulates the formation of clots in contact with blood due to the interaction of the amino groups with the acid groups of blood cells (Hirano et al. 1990). Several works have been reported explaining the interactions of free amino groups of chitosan with plasma proteins or/and blood cells which could cause a thrombogenic or/and a hemolytic response (Hirano et al. 2000).

Chitosan also possesses hypocholesterolemic and hypolipidemic activities (Domard and Domard 2001). Along with these properties, chitosan has antimicrobial, antiviral and antitumor activities and can be used for drug delivery (Ngwuluka et al. 2016), tissue engineering (Deepthi et al. 2016; Ran et al. 2017) and in dentistry (Chang et al. 2017).

Biocompatible chitosan/polyethylene glycol diacrylate (PEGDA) blend films were successfully prepared by Michael addition reaction with different weight ratios as wound dressing materials by Zhang et al. (2008). Mechanical properties and swelling property of chitosan were found to be enhanced after chemical modification, and the films thus obtained also exhibited nontoxicity. These chitosan/PEGDA films have the potential to be used as dressing materials for wounds.

Thakur et al. (2016) found that chitosan montmorillonite polymer composites could be used for the development of various drug delivery systems due to their good compression and release retardant properties in addition to food and packaging applications.

Sodagar et al. (2016) found the bacterial biofilm inhibition in composite containing NPs was significantly higher than the conventional composite resins, and that the properties of composites changed by adding NPs to it. Hence, according to authors, addition of chitosan can decrease plaque accumulation and the mechanical and physical properties remain approximately without changes.

A major problem of chitosan in tissue engineering is its poor solubility in neutral aqueous solutions and organic solvents due to the presence of amino groups and its high crystallinity (Lee and Mooney 2001); therefore, the modification of chitosan is necessary to expand its applications. Chitosan-magnesium-based composite scaffolds for tissue engineering were successfully synthesized by Adhikari et al. (2016). They found that these scaffolds are nontoxic and could provide adequate support for cell growth and proliferation. The chitosan-MgG hybrid scaffold mimicked the extracellular matrix (ECM) of natural tissue physically and chemically, and possessed high surface area, porosity, pore interconnectivity and excellent mechanical stability.

# References

Adhikari U, Rijal NP, Khanal S, Pai D, Sankar J, Bhattarai N (2016) Magnesium incorporated chitosan based scaffolds for tissue engineering applications. Bioact Mater 1:132–139. https://doi.org/10.1016/j.bioactmat.2016.11.003

Al-Naamani L, Dobretsov S, Dutta J (2016) Chitosan-zinc oxide nanoparticle composite coating for active food packaging applications. Innov Food Sci Emerg 38:231–7. https://doi.org/10.1016/j.ifset.2016.10.010

Bhatnagar A, Sillanpää M (2009) Applications of chitin-and chitosan-derivatives for the detoxification of water and wastewater—a short review. Adv Colloid Interface Sci 152:26–38. https://doi.org/10.1016/j.cis.2009.09.003

Chang B, Ahuja N, Ma C, Liu X (2017) Injectable scaffolds: preparation and application in dental and craniofacial regeneration. Mater Sci Eng R: Reports. 111:1–26. https://doi.org/10.1016/j.mser.2016.11.001

Copello GJ, Mebert AM, Raineri M, Pesenti MP, Diaz LE (2011) Removal of dyes from water using chitosan hydrogel/SiO$_2$ and chitin hydrogel/SiO$_2$ hybrid materials obtained by the sol-gel method. J Hazard Mater 186:932–939. https://doi.org/10.1016/j.jhazmat.2010.11.097

Crini G, Badot PM (2008) Application of chitosan, a natural aminopolysaccharide, for dye removal from aqueous solutions by adsorption processes using batch studies: a review of recent literature. Prog Polym Sci 33:399–447. https://doi.org/10.1016/j.progpolymsci.2007.11.001

Deepthi S, Venkatesan J, Kim SK, Bumgardner JD, Jayakumar R (2016) An overview of chitin or chitosan/nano ceramic composite scaffolds for bone tissue engineering. Int J Biol Macromol 93:1338–53. https://doi.org/10.1016/j.ijbiomac.2016.03.041

De Silva RT, Mantilaka MM, Ratnayake SP, Amaratunga GA, de Silva KN (2017) Nano-MgO reinforced chitosan nanocomposites for high performance packaging applications with improved mechanical, thermal and barrier properties. Carbohyd Polym 157:739–747. https://doi.org/10.1016/j.carbpol.2016.10.038

Domard A, Domard M (2001) Chitosan: structure-properties relationship and biomedical applications. Polym biomater 2:187–212. https://doi.org/10.1201/9780203904671.ch9

Gandhi MR, Kousalya GN, Meenakshi S (2011) Removal of copper (II) using chitin/chitosan nano-hydroxyapatite composite. Int J Biol Macromol 48:119–124. https://doi.org/10.1016/j.ijbiomac.2010.10.009

Habiba U, Afifi AM, Salleh A, Ang BC (2017) Chitosan/(polyvinyl alcohol)/zeolite electrospun composite nanofibrous membrane for adsorption of Cr$^{6+}$, Fe$^{3+}$ and Ni$^{2+}$. J Hazard Mater 322:182–194. https://doi.org/10.1016/j.jhazmat.2016.06.028

Harfiz M, Salleh E, Nur S et al (2014) Starch based active packaging film reinforced with empty fruit bunch (EFB) cellulose nanofiber. Procedia Chem 9:23–33. https://doi.org/10.1016/j.proche.2014.05.004

Hassanzadeh P, Kharaziha M, Nikkhah Shin S R, Jin J, He S, Rolandi M (2013) Chitin nanofiber micropatterned flexible substrates for tissue engineering. J Mater Chem B Mater Biol Med 1:4217–4224. https://doi.org/10.1016/j.immuni.2010.12.017

Hirano S, Seino H, Akiyama Y, Nonaka I (1990) Chitosan: a biocompatible material for oral and intravenous administrations. In: Progress in biomedical polymers. Springer, pp 283–290

Hirano S, Zhang M, Nakagawa M, Miyata T (2000) Wet spun chitosan–collagen fibers, their chemical $N$-modifications, and blood compatibility. Biomaterials 21:997–1003. https://doi.org/10.1016/S0142-9612(99)00258-6

Hosseini SF, Rezaei M, Zandi M, Farahmandghavi F (2016) Development of bioactive fish gelatin/chitosan nanoparticles composite films with antimicrobial properties. Food Chem 194:1266–1274. https://doi.org/10.1016/j.foodchem.2015.09.004

Hou H, Zhou R, Wu P, Wu L (2012) Removal of Congo red dye from aqueous solution with hydroxyapatite/chitosan composite. Chem Eng J 211:336–342. https://doi.org/10.1016/j.cej.2012.09.100

Huang GL, Zhang HY, Jeffrey XS, Tim AGL (2009) Adsorption of chromium(VI) from aqueous solutions using cross-linked magnetic chitosan beads. Ind Eng Chem Res 48:2646–2651. https://doi.org/10.1021/ie800814h

Huang Q, Zhang H, Hu S, Li F, Weng W, Chen J, Wang Q, He Y, Zhang W, Bao X (2014) A sensitive and reliable dopamine biosensor was developed based on the Au@ carbon dots–chitosan composite film. Biosens Bioelectron 52:277–280. https://doi.org/10.1016/j.bios.2013.09.003

Huang Y, Yao M, Zheng X, Liang X, Su X, Zhang Y, Lu A, Zhang L (2015) Effects of chitin whiskers on physical properties and osteoblast culture of alginate based nanocomposite hydrogels. Biomacromolecules 16:3499–3507. https://doi.org/10.1021/acs.biomac.5b00928

Jayakumar R, Ramachandran R, Divyarani VV, Chennazhi KP, Tamura H, Nair SV (2011a) Fabrication of chitin-chitosan/nano TiO$_2$-composite scaffolds for tissue engineering applications. Int J Biol Macromol 48:336–344. https://doi.org/10.1016/j.ijbiomac.2010.12.010

Jayakumar R, Ramachandran R, Sudheesh Kumar PT, Divyarani VV, Chennazhi KP, Tamura H, Nair SV (2011b) Fabrication of chitin-chitosan/nano ZrO$_2$ composite scaffolds for tissue engineering applications. Int J Biol Macromol 49:274–280. https://doi.org/10.1016/j.ijbiomac.2011.04.020

Karthik R, Meenakshi S (2015) Synthesis, characterization and Cr(VI) uptake study of polyaniline coated chitin. Int J Biol Macromol 72:235–242. https://doi.org/10.1016/j.ijbiomac.2014.08.022

Kumar PTS, Abhilash S, Sreeja V, Tamura H, Manzoor K, Nair SV, Jayakumar R (2010) Development of novel chitin/nanosilver composite scaffolds for wound dressing applications. J Mater Sci Mater Med 21:807–813. https://doi.org/10.1007/s10856-009-3877-z

Lee KY, Mooney DJ (2001) Hydrogels for tissue engineering. Chem Rev 101:1869–80. https://doi.org/10.1021/cr000108x

Li X, Li X, Ke B et al (2011) Cooperative performance of chitin whisker and rectorite fillers on chitosan films. Carbohydr Polym 85:747–752. https://doi.org/10.1016/j.carbpol.2011.03.040

Liu P, Sehaqui H, Tingaut P, Wichser A, Oksman K, Mathew AP (2014) Cellulose and chitin nanomaterials for capturing silver ions ($Ag^+$) from water via surface adsorption. Cellul 21:449–461. https://doi.org/10.1007/s10570-013-0139-5

Mangalathillam S, Rejinold NS, Nair A, Lakshmanan VK, Nair SV, Jayakumar R (2012) Curcumin loaded chitin nanogels for skin cancer treatment via the transdermal route. Nanoscale 4:239–250. https://doi.org/10.1039/c1nr11271f

Ngah WW, Teong LC, Hanafiah MA (2011) Adsorption of dyes and heavy metal ions by chitosan composites: a review. Carbohydr polym 283(4):1446–1456. https://doi.org/10.1016/j.carbpol.2010.11.004

Nasri-Nasrabadi B, Mehrasa M, Rafienia M, Bonakdar S, Behzad T, Gavanji S (2014) Porous starch/cellulose nanofibers composite prepared by salt leaching technique for tissue engineering. Carbohydr Polym 108:232–238

Ngwuluka NC, Ochekpe NA, Aruoma OI (2016) Functions of bioactive and intelligent natural polymers in the optimization of drug delivery. In: Industrial applications for intelligent polymers and coatings. Springer International Publishing, pp 165–184

Olivera S, Muralidhara HB, Venkatesh K, Guna VK, Gopalakrishna K, Kumar Y (2016) Potential applications of cellulose and chitosan nanoparticles/composites in wastewater treatment: a review. Carbohydr Polym 153:600–618. https://doi.org/10.1016/j.carbpol.2016.08.017

Pantaleone D, Yalpani M, Scollar M (1992) Unusual susceptibility of chitosan to enzymic hydrolysis. Carbohydr Res 237:325–332. https://doi.org/10.1016/S0008-6215(92)84256-R

Qin Y, Zhang S, Yu J, Yang J, Xiong L, Sun Q (2016) Effects of chitin nano-whiskers on the antibacterial and physicochemical properties of maize starch films. Carbohydr Polym 147:372–378. https://doi.org/10.1016/j.carbpol.2016.03.095

Raj V, Prabha G (2016) Synthesis, characterization and in vitro drug release of cisplatin loaded Cassava starch acetate – PEG/gelatin nanocomposites. J Assoc Arab Univ Basic Appl Sci 21:10–16. https://doi.org/10.1016/j.jaubas.2015.08.001

Ran J, Jiang P, Sun G, Ma Z, Hu J, Shen X, Tong H (2017) Comparisons among Mg, Zn, Sr and Si doped nano-hydroxyapatite/chitosan composites for load-bearing bone tissue engineering applications. Mater Chem Front 1:900–910. https://doi.org/10.1039/c6qm00192k

Rhim JW, Park HM, Ha CS (2013) Bio-nanocomposites for food packaging applications. Prog Polym Sci 38:1629–1652. https://doi.org/10.1016/j.progpolymsci.2013.05.008

Salaberria AM, Diaz RH, Labidi J, Fernandes SCM (2015a) Role of chitin nanocrystals and nanofibers on physical, mechanical and functional properties in thermoplastic starch films. Food Hydrocoll 46:93–102. https://doi.org/10.1016/j.foodhyd.2014.12.016

Salaberria AM, Labidi J, Fernandes SCM (2015b) Different routes to turn chitin into stunning nano-objects. Eur Polym J 68:503–515. https://doi.org/10.1016/j.eurpolymj.2015.03.005

Shan C, Yang H, Han D, Zhang Q, Ivaska A, Niu L (2010) Graphene/AuNPs/chitosan nanocomposites film for glucose biosensing. Biosens Bioelectron 25:1070–1074. https://doi.org/10.1016/j.bios.2009.09.024

Shankar S, Reddy JP, Rhim JW, Kim HY (2015) Preparation, characterization, and antimicrobial activity of chitin nanofibrils reinforced carrageenan nanocomposite films. Carbohydr Polym 117:468–475. https://doi.org/10.1016/j.carbpol.2014.10.010

Shen C, Shen Y, Wen Y, Wang H, Liu W (2011) Fast and highly efficient removal of dyes under alkaline conditions using magnetic chitosan-Fe (III) hydrogel. Water Res 45:5200–5210. https://doi.org/10.1016/j.watres.2011.07.018

Shukla SK, Mishra AK, Arotiba OA, Mamba BB (2013) Chitosan-based nanomaterials: a state-of-the-art review. Int J Biol Macromol 59:46–58. https://doi.org/10.1016/j.ijbiomac.2013.04.043

Sivakami MS, Gomathi T, Venkatesan J, Jeong HS, Kim SK, Sudha PN (2013) Preparation and characterization of nano chitosan for treatment wastewaters. Int J Biol Macromol 57:204–212. https://doi.org/10.1016/j.ijbiomac.2013.03.005

Sodagar A, Bahador A, Jalali YF, Gorjizadeh F, Baghaeian P (2016) Effect of Chitosan Nanoparticles Incorporation on Antibacterial Properties and Shear Bond Strength of Dental Composite Used in Orthodontics. Iran J Ortho. https://doi.org/10.17795/ijo-7281

Souza AC, Goto GEO, Mainardi JA, Coelho ACV, Tadini CC (2013) Cassava starch composite films incorporated with cinnamon essential oil: antimicrobial activity, microstructure, mechanical and barrier properties. LWT - Food Sci Technol 54:346–352. https://doi.org/10.1016/j.lwt.2013.06.017

Tang H, Chang C, Zhang L (2011) Efficient adsorption of $Hg^{2+}$ ions on chitin/cellulose composite membranes prepared via environmentally friendly pathway. Chem Eng J 173:689–697. https://doi.org/10.1016/j.cej.2011.07.045

Thakur G, Singh A, Singh I (2016) Chitosan-montmorillonite polymer composites: formulation and evaluation of sustained release tablets of aceclofenac. Sci Pharm 84:603–617. https://doi.org/10.3390/scipharm84040603

Tran HV, Tran LD, Nguyen TN (2010) Preparation of chitosan/magnetite composite beads and their application for removal of Pb(II) and Ni(II) from aqueous solution. Mater Sci Eng, B 30:304–310. https://doi.org/10.1016/j.msec.2009.11.008

Tzoumaki MV, Moschakis T, Biliaderis CG (2011) Mixed aqueous chitin nanocrystal-whey protein dispersions: microstructure and rheological behaviour. Food Hydrocoll 25:935–942. https://doi.org/10.1016/j.foodhyd.2010.09.004

Usami Y, Okamoto Y, Takayama T, Shigemasa Y, Minami S (1998) Chitin and chitosan stimulate canine polymorphonuclear cells to release leukotriene $B_4$ and prostaglandin $E_2$. J Biomed Mater Res 42:517–522. https://doi.org/10.1002/(SICI)1097-4636(19981215)42:4%3c517:AID-JBM6%3e3.0.CO;2-U

Wang L, Wang A (2007) Adsorption characteristics of Congo Red onto the chitosan/montmorillonite nanocomposite. J Hazard Mater 147:979–985. https://doi.org/10.1016/j.jhazmat.2007.01.145

Wysokowski M, Behm T, Born R, Bazhenov VV, Meißner H, Richter G, Szwarc-Rzepka K, Makarova A, Vyalikh D, Schupp P, Jesionowski T (2013a) Preparation of chitin–silica composites by in vitro silicification of two-dimensional Ianthella basta demosponge chitinous scaffolds under modified Stöber conditions. Mater Sci Eng, C 33:3935–3941. https://doi.org/10.1016/j.msec.2013.05.030

Wysokowski M, Motylenko M, Stöcker H, Bazhenov VV, Langer E, Dobrowolska A, Czaczyk K, Galli R, Stelling AL, Behm T, Klapiszewski L (2013b) An extreme biomimetic approach: hydrothermal synthesis of β-chitin/ZnO nanostructured composites. J Mater Chem B 1:6469–6476. https://doi.org/10.1039/c3tb21186j

Zargar V, Asghari M, Dashti A (2015) A review on chitin and chitosan polymers: structure, chemistry, solubility, derivatives, and applications. Chem Bio Eng Rev 2:204–226. https://doi.org/10.1002/cben.201400025

Zhang X, Yang D, Nie J (2008) Chitosan/polyethylene glycol diacrylate films as potential wound dressing material. Int J Biol Macromol 43:456–462. https://doi.org/10.1016/j.ijbiomac.2008.08.010

Zhou L, Jin J, Liu Z, Liang X, Shang C (2011) Adsorption of acid dyes from aqueous solutions by the ethylenediamine-modified magnetic chitosan nanoparticles. J Hazard Mater 185:1045–52. https://doi.org/10.1016/j.jhazmat.2010.10.012

# Chapter 5
# Conclusion

In this book, an attempt has been made to understand the importance and characteristics of polysaccharides like starch, chitin and chitosan with special thrust on thermal, morphological and tensile characteristics. They are attractive biomaterials for a multitude of potential applications in a diverse range of fields. Starch is considered to be one of the most promising natural polymer candidates available for the development of biodegradable materials. Chitin as a polymer, as well as a reinforcing agent, has an excellent potential for applications which still needs to be developed. Chitosan can be used as a natural polymer matrix as well as nanofillers for the fabrication of composites because of its interesting biological properties. The growing list of literature studying polysaccharides, mainly starch, chitin and chitosan, is a clear indication of their evolution. Practical applications of such biomaterials in industrial technology require a favorable balance between the expected performance of the composite materials and their cost. Research and development investments must be made in science and engineering fields that will fully determine the properties and characteristics of polysaccharides and new technologies should be developed so that industry can produce advanced and cost competitive polysaccharide composite products. This means that there are still significant scientific and technological challenges to take up.

© The Author(s), under exclusive licence to Springer Nature Switzerland AG 2019
M. S. Thomas et al., *Starch, Chitin and Chitosan Based Composites
and Nanocomposites*, Biobased Polymers,
https://doi.org/10.1007/978-3-030-03158-9_5

Printed in the United States
By Bookmasters